All the Math You'll Ever Need

A Self-Teaching Guide
Revised Edition

Steve Slavin

John Wiley & Sons, Inc.

New York • Chichester • Weinheim • Brisbane • Singapore • Toronto

This book is printed on acid-free paper. ♾

Published by John Wiley & Sons, Inc.
Published simultaneously in Canada.

Library of Congress Cataloging-in-Publication Data:
Slavin, Steve.
All the math you'll ever need : a self-teaching guide / Steve Slavin. — Rev. ed.
p. cm.
Includes index.
ISBN 0-471-31751-9 (pbk. : alk. paper)
ISBN 0-471-38241-8 (cloth : alk. paper)
1. Mathematics. I. Title.
QA39.2.S56 1999
510—dc21 98-37205

Printed in the United States of America

22

Contents

How to Use This Book vii

Acknowledgments x

1 Getting Started 1

2 Simple Multiplication 4
What Is Multiplication?, 4
The Multiplication Table, 5
Learning the Multiplication Table, 5

3 Long Multiplication 10
Testing Your Knowledge, 10
About Long Multiplication, 11
Practice Exercises with Two Digits, 13
Multiplying Three Digits, 14
Four-Digit Multiplication, 15
Word Problems with Multiplication, 17

4 Division 19
Short Division, 19
Long Division, 22
Word Problems with Division, 25

5 Fast Multiplication and Division 28
Fast Multiplication, 28
Fast Division, 30
How Are You Doing So Far?, 31

6 Multiplying and Dividing with Decimals 32
Multiplying with Decimals, 32
Dividing with Decimals, 34

7 Converting Fractions into Decimals and Decimals into Fractions 36
Converting Fractions into Decimals, 36
Converting Decimals into Fractions, 38
Reducing Fractions to Their Lowest Denominators, 39

8 Adding, Subtracting, Multiplying, and Dividing Fractions 41
Adding Fractions, 41
Subtracting Fractions, 44
Multiplying by Fractions, 45
Dividing by Fractions, 47
Cancelling Out, 49

9 Advanced Word Problems 51
Words and Fractions, 51
Word Problems with Money, 53

10 Percentages 56
Fractions, Decimals, and Percentages, 56
Percentage Changes, 59
Fast Percentage Changes, 60
Percentage Distribution, 62
Tipping, 65

11 Negative Numbers 67
What Are Negative Numbers?, 67
Adding Negative Numbers, 68
Subtracting Negative Numbers, 68
Multiplying Negative Numbers, 69
Division with Negative Numbers, 71

12 The Isolation of x 72

The Care and Treatment of Equations, 72
Isolating x, 72
Addition and Subtraction with x, 73
Multiplication and Division with x, 74
Decimals with x, 76
Negative Numbers, 77
Combination Problems with x, 79

13 Exponents and Square Roots 83

Exponents: The Powers of x, 83
Square Roots, 85
Advanced Problems, 87

14 Ratios and Proportions 89

Ratios, 89
Proportions, 91

15 Finding the Areas of Rectangles and Triangles 95

Areas of Rectangles, 95
When We Use Area, 98
Perimeters of Rectangles, 100
Areas of Triangles, 102

16 Finding the Circumferences and Areas of Circles 105

Circumferences of Circles, 105
The Area of a Circle, 108

17 Rate, Time, and Distance Problems 112

The Magic Formula, 112
The Terms, 112
Finding Distance, 113
Finding Rate, 115
Finding Time, 120
Rate, Time, and Distance Problems, 124
Speed Limit Problems, 126

18 Finding x: Advanced Problems 131

Age Problems, 131

Finding the Numbers, 134
Nut Problems, 137

19 Interest Rates 141
Simple Interest, 141
Compound Interest, 143
Doubling Time: The Rule of 70, 146
Discounting, 148
The True Rate of Interest, 150

20 Big Numbers 153
Thousands and Millions, 153
Billions and Trillions, 155
Quadrillions, Quintillions, and Even Bigger Numbers, 158
Multiplying Big Numbers, 159
Dividing Big Numbers, 160

21 A Taste of Statistics 163
The Mean, the Median, and the Mode, 163

22 Personal Finances 166
Mark-Down Problems, 166
Sales Tax Problems, 170
Credit Card Problems, 173
Figuring Your Federal Income Tax, 176
Mortgage Interest and Taxes, 179

23 Business Math 183
Commissions, 183
Mark-Ups, 186
Discounting from List Price, 188
Quantity Discounts, 191
2/10 n/30, 193
Chain Discounts, 197
Profit, 199

24 Review 206
Where To from Here?, 227

Index 229

How to Use This Book

How long has it been since you've worked out math problems? When I was growing up in Brooklyn in the 1940s, my mother bought fruit and vegetables from Mr. Levine, who added sums on a paper bag. Today we've got cashiers in fast-food restaurants who've got pictures of each item right on the cash register, so they don't even have to punch in the dollars and cents each item costs. No wonder so many of us have forgotten how to use numbers.

This book is organized by chapter with periodic self-tests throughout each chapter. Their purpose is to make sure you comprehend material before moving on. If you find that you have made an error, look back at the preceding material to make sure you understand the correct answer. The information is arranged so that it builds on what comes before. To fully understand the information at the end of a chapter, you must first have completed all of the preceding self-tests.

Before you even start, please put away your pocket calculator. That's right! You're going to be able to go through this entire book, working out hundreds of problems, without once having to use a calculator.

Admit it—you have an addiction. You have a pocket calculator habit. And just look at how that dependency has affected your mathematical skills. The more you came to depend on your calculator, the harder it became to do simple multiplication and division, either in your head or on paper. Do you want to go through the rest of your life in this condition, or do you want to break out of this vicious circle?

Recently, my friend Donna bought $1,536.40 worth of groceries at the supermarket. Usually her bills are somewhat lower, but once a year she's entitled to a 10% discount. That day, however, she was out of luck, because the cashier had forgotten to bring her calculator to work. In exasperation Donna blurted out, "Ten percent of $1,536.40 is $153.64." The cashier was so amazed that she called over all her fellow cashiers *and* the store manager to meet this great math whiz.

The trick to learning math is moving ahead at just the right pace. The format of this book lends itself to proper pacing. When you're going too slowly, you'll say to yourself, "This stuff is so easy—I'm getting bored." So you'll be able to skip a few sections and move on to new material. But when you find yourself pounding your fists against the wall and despairing of ever learning math, that may mean you've been moving ahead a bit too quickly.

Beginning with chapter 3, every chapter has at least two self-tests. If you feel that you don't need to read a particular chapter, you may want to take the self-tests anyway. These provide not only a quick review of the subject matter covered in the chapter, but also a good way of gauging what you already know.

Should you find, on the other hand, that you're having trouble doing a certain type of problem, it will be made clear to you that you need to review an earlier section. For example, no one can do simple division without knowing the multiplication table. So everyone who gets stuck at this point will be sent back to learn that table once and for all. Once that's accomplished, it will be clear sailing through the next few chapters.

This book provides a fast-paced review of arithmetic and elementary algebra, with a smattering of statistics thrown in. It is intended to refresh the memory of the high school or college graduate, the person whose mathematical thought processes have atrophied over the years.

The main emphasis here is on getting you to rely on your own mathematical skills. After completing this book, you will have risen above the ranks of the mathematically illiterate.

Don't be ashamed to carry this book around with you. Sure, people will stop you in the street and ask you questions. No, they won't ask you to multiply 798,999 by 516,204. They'll probably want to know where *they* can get a copy of the book.

So you'll meet a lot of people and become very popular. And as your math skills come back to you, you'll find your whole life has changed. You'll show off in restaurants with your friends; when the check comes, you'll calculate everyone's share in your head. But the main thing you'll find is that you'll have lost your fear of numbers. New horizons will open up to you. And all of this will happen because you have overcome your dependency on your pocket calculator.

This book will do the same thing for you that the Charles Atlas course did for 97-pound weaklings. No longer will you be intimidated trying to calculate tips. No longer will you need to whip out your pocket calculator to do simple arithmetic. And you won't have to wait months to see tangible results. You won't even have to wait weeks. In just a few days your friends and colleagues will notice your new mathematical muscles. That's right! No one will ever dare to call you a mathematical illiterate again! And when you walk in the street, your neighbors will look at you with new respect. So don't delay another minute. Turn to chapter 1 and just watch those brain cells start to grow.

Acknowledgments

Many thanks are due, so I'd like to name names. My longtime editor at Wiley, Judith McCarthy, made hundreds of suggestions to improve and update the book. Authors often hate to change even a word, but Judith's editing has made this second edition a much smoother read. Claire McKean did a thorough copyedit, catching dozens of errors that made it through the first edition, and Benjamin Hamilton supervised the production of the book from copyediting through page proofs.

I owe a large debt of gratitude to my family, especially to my nephews, Jonah and Eric Zimiles. Jonah provided me with a blow-by-blow critique of the strengths and weaknesses of my previous book, *Economics: A Self-Teaching Guide* (Wiley, 1988), on which I was able to build while writing *this* book. And Eric, after having read that book, recognized its format lent itself best to my writing style and encouraged me to write another book. Eric's daughters, Eleni, 11, Justine, 7, and Sophie, 5, have contributed to the new edition by helping me with my math whenever I happened to get stuck.

My father, Jack, a retired math teacher, provided inspiration of another kind. As the oldest living academic perfectionist, he upholds such unattainable standards that one cannot help but feel tolerance for one's own shortcomings and those of just about everyone else. And finally, I wish to thank my sister, Leontine Temsky, for her rationality and common sense in the most uncommon and irrational of times.

1 Getting Started

Tens of millions of Americans are mathematically illiterate. They have trouble doing even third- and fourth-grade arithmetic. And many of these people are college graduates. One cannot help but wonder how so many people managed to get so far in school without having mastered basic arithmetic, let alone simple algebra.

Math phobia seems to have spread across the country. Whenever I mention more than one or two numbers in the same sentence, I notice that my listeners' eyes begin to glaze over. Like counting sheep, just the mention of a couple of numbers is a surefire way to put an audience to sleep.

As an economics professor, one of the strangest phenomena I've encountered is that so many college freshmen are unable to multiply and divide. After all, they studied that back in elementary, middle, or junior high school; since then they've had a couple of years of high school algebra.

So how come they can't do third- and fourth-grade arithmetic? And even more to the point, what can be done to make up for this deficiency?

That is how I came to write this book. I needed results, and I needed them fast, because you can't understand economics if you can't crunch a few numbers.

But you don't need to be an economics student just to deal with numbers. We all have to deal with numbers *sometime*—to balance our checkbooks, to do our income tax, and to decide whether we want a fixed- or variable-interest mortgage.

Like it or not, numbers are an important part of our lives, and it is time they stopped intimidating us.

There's an apocryphal story that made the rounds back in the 1950s. A big Cadillac pulls up outside Grossinger's, a fancy resort located in the Catskill Mountains. Out steps a very well dressed couple. And right behind them comes their chauffeur carrying a five- or six-year-old boy.

Jenny Grossinger herself rushes out to greet them. She tells them how sorry she is about their little boy.

"Sorry?" replies the boy's mother. "Why are you sorry?"

"Because he can't walk."

"Of *course* he can walk," replies the mother. "Thank God he doesn't *have* to."

What does this have to do with mathematical illiteracy? Just as that little boy who never has to walk will lose the use of his legs, so too will the person who doesn't use his or her mathematical abilities lose them. When you rely on something else to help you, in this case your pocket calculator, your skills grow rusty.

Clearly the villain is the pocket calculator. That's right! Why bother to multiply and divide when your calculator will save you the trouble? In fact, why bother to do any arithmetic at all when all you need to do is press a couple of buttons and read the answer shown in the window?

So what we'll be doing here is going back to basics. Since nearly everyone who can count past 10 knows how to add and subtract, we'll start off with the multiplication table. You'll need to memorize it. If you need to go over addition and subtraction, you need to consult an even more basic text, such as *Quick Arithmetic* by Robert A. Carmen and Marilyn J. Carmen (Wiley, 1987).

If you don't know the multiplication table, you make an awful lot of mistakes in multiplication *and* division. And even if you once knew the multiplication table, if you came to depend on a calculator, your skills at multiplication and division have eroded.

In fact, what has probably happened is that you have developed a math phobia. Numbers make a lot of people nervous. I often hear, "When I see numbers, I freeze . . . or panic . . . or just give up."

So what it all comes down to is learning the multiplication table, using that knowledge to multiply and divide, and then going on to even bigger and better things.

In this text, the use of complex formulas is purposely avoided. Although such formulas have an honored place in mathematics, they clearly don't belong in a book that is read by people who are trying to get over their fear of numbers.

The use of Greek letters will also be eliminated in this book. A simple *x* for an unknown will generally suffice.

Finally, the use of technical terms is avoided whenever possible. You will rarely encounter such terms as "numerator" or "denominator" in this book, and the term "exponent" will only be used three times on the first page of chapter 13—it will not be mentioned again after that. There are no quadratic formulas, logarithmic tables, integrals, or derivatives, and there are only a handful of very simple graphs. As long as you learn the math, we'll be happy to skip the vocabulary lesson.

Before we get too carried away, a word of caution is in order. This is not quite the mathematical equivalent of Lourdes. You will not be asked to throw away your calculator as some have thrown away their crutches and wheelchairs. Just put it in a safe place for now, to be taken out and used only on special occasions. A calculator is most effectively used for three tasks: (1) to do calculations that need to be done rapidly, (2) to do repetitive calculations, and (3) to do sophisticated calculations that would take up to a couple of hours to do without a calculator.

The trick is to use our calculators for these specific tasks and not for arithmetic functions that we can do in our heads. Just as many suburbanites drive everywhere—even to the corner to buy a paper—so do many of us depend on our calculators to do simple addition, subtraction, multiplication, and division. So put away your calculator and start using your innate mathematical ability.

2 Simple Multiplication

Once you can multiply without using a calculator, you will find the entire world of mathematics opened up to you. The point of this chapter is to make sure you know the entire multiplication table from 1×1 to 10×10.

If you are sure that you do, then go directly to Table 2.1 (page 7) and fill it in completely. Then use Table 2.2 (page 8) to check your work. If you get everything right, then go on to chapter 3. But if you get even two wrong, return to frame 1.

1 What Is Multiplication?

Multiplication is addition. For instance, how much is 4×3? You know it's 12 because you searched your memory for that multiplication fact. There's nothing wrong with that.

Another way to calculate 4×3 is to add them:

$3 + 3 + 3 + 3 = 12$ or $4 + 4 + 4 = 12$

Solve the problem 5×7. Maybe you know it's 35. You could always do this:

$7 + 7 + 7 + 7 + 7 = 35$ or $5 + 5 + 5 + 5 + 5 + 5 + 5 = 35$

We do multiplication instead of addition because it's shorter—sometimes much shorter. Suppose you needed to multiply 78×95. If you set this up as an addition problem, you'd be calculating for an hour.

Let's set this up as a regular multiplication problem and take the very first step toward solving it:

$$\begin{array}{r} 95 \\ \underline{\times 78} \end{array}$$

The first set of numbers we'd multiply would be 8×5. Then we'd multiply 8×9. Next would come 7×5 and 7×9. That's a lot easier than writing out a column of 78 95s and adding them up.

As you can see, a long multiplication problem can be broken down into a series of simple multiplication problems. In this chapter, we will concentrate on these short, or simple, multiplication problems (nothing past 10×10). In the next chapter, we'll tackle longer multiplication.

2 The Multiplication Table

The bare bones are laid out in Table 2. 1. You'll be filling in the blanks.

The table shown here goes from 1×1 to 10×10. When I learned the table way back in the third grade, shortly after the Civil War, we had to learn up to 12×12. But I'll let you off the hook. After all, how many times do you need to multiply 12×11?

How well do you know the table? There's only one way to find out. Fill it in. Multiply each number in the vertical column by each number in the horizontal row. I've done a few to get you started.

Once you've finished, use Table 2.2 to check your work. Circle any wrong answers. The problems most commonly missed are 9×6 and 8×7.

If you got everything right, you may advance to the next chapter. If you got just one or two of these wrong, then you'll need to go over them several times, until you're sure you know them.

What if you just don't know the multiplication table at all? There are no shortcuts. Just keep reading. And by the time you've finished this chapter, you'll be able to do the multiplication table in your sleep.

3 Learning the Multiplication Table

Let's return to a concept discussed in frame 1—multiplication as addition. Fill in the missing numbers of these five series:

TABLE 2.1

BLANK MULTIPLICATION TABLE

Please fill in completely. If you're not too sure of yourself, do it in pencil first.

	1	2	3	4	5	6	7	8	9	10
1	1	2	3	4	5	6	7	8	9	10
2	2	4	6	8	10	12	14	16	18	20
3	3	6	9	12	15	18	21	24	27	30
4	4	8	12	16	20	24	28	32	36	40
5	5	10	15	20	25	30	35	40	45	50
6	6	12	18	24	30	36	42	48	54	60
7	7	14	21	28	35	42	49	56	63	70
8	8	16	24	32	40	48	56	64	72	80
9	9	18	27	36	45	54	63	72	81	90
10	10	20	30	40	50	60	70	80	90	100

1	2	3	4	5	6	7	8	9	10
2	4	6	8	10	12	14	16	18	20
3	6	9	12	15	18	21	24	27	30
4	8	12	16	20	24	28	32	36	40
5	10	15	20	25	30	35	40	45	50

Now check your work:

1	2	3	4	5	6	7	8	9	10
2	4	6	8	10	12	14	16	18	20
3	6	9	12	15	18	21	24	27	30

TABLE 2.2

COMPLETED MULTIPLICATION TABLE

	1	2	3	4	5	6	7	8	9	10
1	1	2	3	4	5	6	7	8	9	10
2	2	4	6	8	10	12	14	16	18	20
3	3	6	9	12	15	18	21	24	27	30
4	4	8	12	16	20	24	28	32	36	40
5	5	10	15	20	25	30	35	40	45	50
6	6	12	18	24	30	36	42	48	54	60
7	7	14	21	28	35	42	49	56	63	70
8	8	16	24	32	40	48	56	64	72	80
9	9	18	27	36	45	54	63	72	81	90
10	10	20	30	40	50	60	70	80	90	100

4 8 12 16 20 24 28 32 36 40

5 10 15 20 25 30 35 40 45 50

If you got everything right, then you're ready for the next set. If not, you'll need to go over the ones you got wrong until you've mastered them.

We'll begin with sixes. Even if you need to count on your fingers, that's okay.

6 12 18 24 30 36 42 48 54 60

Next we'll do sevens:

7 14 21 27 35 42 49 56 63 70

Now check your work:

6	12	18	24	30	36	42	48	54	60
7	14	21	18	35	42	49	56	63	70

How did you do? If you got them all right, go on to the next set. If not, continue working on those you missed.

We're ready to do eights:

8	16	24	____	____	____	____	____	____	____

Now try nines:

9	18	27	____	____	____	____	____	____	____

Finally, fill in the missing numbers for tens:

10	20	30	____	____	____	____	____	____	____

Now check your work:

8	16	24	32	40	48	56	64	72	80
9	18	27	36	45	54	63	72	81	90
10	20	30	40	50	60	70	80	90	100

Need more practice? Return to the beginning of frame 3 and redo all the exercises. Fill in Table 2.3 and check your work against Table 2.2.

If you're still getting some of these wrong, then you're ready for flash cards. Write the problem on one side of an index card and the answer on the other side. Practice them until you have memorized the correct answer to each of these problems.

Don't get discouraged. Once you've mastered these problems, you'll own them for the rest of your life. The only catch is that you'll have to keep using them. It's like learning how to type. Once you know the multiplication table, your math skills will keep improving.

As you get used to working with numbers, you'll find things getting much easier. Numbers will seem a lot less intimidating, and you'll find yourself gaining confidence. It's really a cumulative process that began

TABLE 2.3

ANOTHER BLANK MULTIPLICATION TABLE

	1	2	3	4	5	6	7	8	9	10
1	1	2								
2	2	4								
3			9	12						
4										
5										
6										
7										
8										
9										
10										

when you discarded your calculator and learned or relearned the multiplication table.

We're now ready to apply what we've learned here by using the multiplication table in the next chapter. Remember, if you don't use it, you'll lose it.

3 Long Multiplication

We'll be doing some longer, more complex multiplication problems in this chapter. But first you should determine whether you really should be reading this chapter.

Testing Your Knowledge

We need to separate the advanced multipliers from the nonmultipliers. If you don't know your multiplication table, you're not yet ready for this chapter. Please go back to the beginning of chapter 2. On the other hand, you may already know how to calculate more complex multiplication problems. There's only one way to find out—sink or swim. Please work out these three problems:

$$\begin{array}{r} 89 \\ \underline{\times 57} \end{array} \qquad \begin{array}{r} 195 \\ \underline{\times 473} \end{array} \qquad \begin{array}{r} 7064 \\ \underline{\times 3978} \end{array}$$

Solutions:

$$\begin{array}{r} 89 \\ \underline{\times 57} \\ 623 \\ \underline{445} \\ 5073 \end{array} \qquad \begin{array}{r} 195 \\ \underline{\times 473} \\ 585 \\ 1365 \\ \underline{780} \\ 92235 \end{array} \qquad \begin{array}{r} 7064 \\ \underline{\times 3978} \\ 56512 \\ 49448 \\ 63576 \\ \underline{21192} \\ 28100592 \end{array}$$

Did you notice that our answers had no commas? Certainly 28100592 looks better this way: 28,100,592. But we've left out the commas so our numbers would align vertically. To avoid mistakes, when *you* do long multiplication problems, make sure that *your* figures are aligned. And since you're not typesetting a book, be sure to put in all the commas.

If you got just the first problem right, then you're definitely in the right chapter. If all three problems were correct, then proceed to frame 6. Don't worry, we won't be covering any new material before then.

If you answered two out of three problems correctly, keep reading for the next few frames, work out a few more problems, and then, if you feel you really know how to do long multiplication, you may want to skip to frame 6.

2 About Long Multiplication

Long multiplication is just simple multiplication combined with addition. Let's analyze the first problem. When we multiply 89×57, we're multiplying 89×7, then 89×5, and then adding the two products.

Let's walk through all the steps:

1. $$\begin{array}{r} 89 \\ \underline{\times 57} \end{array}$$

2. $7 \times 9 = 63$

3. Write down the 3 and carry the 6.

4. $$\begin{array}{rl} 89 & \text{carry 6} \\ \underline{\times 57} & \\ 3 & \end{array}$$

5. $7 \times 8 = 56$

6. $56 + \text{the carried } 6 = 62$

7. Write down 62.

8. $$\begin{array}{r} 89 \\ \underline{\times 57} \\ 623 \end{array}$$

9. $5 \times 9 = 45$

10. Write down the 5 and carry the 4.

11.
```
  89 carry 4
 ×57
 623
 5
```

12. 5 × 8 = 40

13. 40 + the carried 4 = 44

14. Write down 44.

15.
```
   89
  ×57
  623
 445
 5073
```

16. Add 623 + 445 = 5073.

Notice how straight my columns are. One number right under another. I do have a slight advantage, since the typesetter did this for me. But it should be obvious that you can avoid a lot of mistakes if you keep your columns straight.

Also notice that we need to indent when we go from one row of numbers to the next. When we're ready for the second row, the 5 goes under the 2 of 623. Each new row requires an indentation.

Why put the 5 under the 2? Why not write it under the 3 like this?

```
  89
 ×57
 623
 445
```

When we start a new row, we always indent one place to the left because that number is ten times the value of a number one place to the right. In this case, we've placed that 5 in the tens column, making it 50. When we multiply, we are rarely, if ever, conscious that when we indent, we are arranging our numbers in columns of units, tens, hundreds, and thousands. But when we get to the answer, we read it as five thousand and seventy-three (five thousands, zero hundreds, five tens, and three units).

3 Practice Exercises with Two Digits

For further practice work out the next two problems and check the solutions:

64	59
×94	×30

Solutions:

64	59
×94	×30
256	00
576	177
6016	1770

How did you do? If you got these two right, go on to frame 4. If not, then let's spend some time going over both problems.

The first one is straightforward: $4 \times 4 = 16$, carry the 1; $4 \times 6 = 24$; $24 + 1 = 25$. So we have 256 in the first row.

We begin the second row by multiplying $9 \times 4 = 36$. Place the 6 directly under the 5 of 256. We must always indent by one place when we start a new row. So we write down the 6 of 36 and carry the 3. Next we multiply $9 \times 6 = 54 + 3 = 57$. This gives us 576 in the second row. Then we add.

Now we'll examine the second problem. How much is zero times any number? It's zero. Since zero has no value, there's nothing wrong with adding a few zeros to the left of any number. (You may add zeros to the right of the number zero without changing its value, but if you added zero to the right of a whole number, such as 17, for example, you'd be multiplying this number by 10, ending up with 170.)

So we'll put two zeros in the first row. Remembering to indent, we multiply $3 \times 9 = 27$, placing the 7 in the second row and carrying the 2. Then $3 \times 5 = 15 + 2 = 17$, giving us 177 in the second row. The rest is addition. In this way, complex multiplication is broken down into a series of simple multiplication and addition steps.

Solve these problems:

80	63
×97	×50

Solutions:

$$\begin{array}{r} 80 \\ \times 97 \\ \hline 560 \\ 720 \\ \hline 7760 \end{array} \qquad \begin{array}{r} 63 \\ \times 50 \\ \hline 00 \\ 315 \\ \hline 3150 \end{array}$$

If you got these right, go on to frame 4; if not, check to be sure you know the multiplication table. If you don't, you'll need to go back to the beginning of chapter 2. If, on the other hand, you do know the table, then begin rereading this chapter.

4 Multiplying Three Digits

Try to solve these two problems:

$$\begin{array}{r} 672 \\ \times 305 \\ \hline \end{array} \qquad \begin{array}{r} 580 \\ \times 941 \\ \hline \end{array}$$

204960 545780

Solutions:

$$\begin{array}{r} 672 \\ \times 305 \\ \hline 3360 \\ 000 \\ 2016 \\ \hline 204960 \end{array} \qquad \begin{array}{r} 580 \\ \times 941 \\ \hline 580 \\ 2320 \\ 5220 \\ \hline 545780 \end{array}$$

If you got both problems right, go on to frame 5; if not, then just keep reading.

Let's go over the first problem. The first row is straightforward: $5 \times 2 = 10$; write down 0 and carry 1; $5 \times 7 = 35 + 1 = 36$; write down 6 and carry the 3; $5 \times 6 = 30 + 3 = 33$.

The next row is all zeros. The third row: $3 \times 2 = 6$; $3 \times 7 = 21$; write down 1 and carry 2; $3 \times 6 = 18 + 2 = 20$. Then add.

Here's an altemate way of doing this problem:

$$\begin{array}{r} 672 \\ \times 305 \\ \hline 3360 \\ 20160 \\ \hline 204960 \end{array}$$

When you're at ease with numbers, you can do it this way. But if you're not yet that much at ease, continue writing out all the zeros.

The second problem is a little easier. The first row is easily obtained. In the second row, $4 \times 0 = 0$; $4 \times 8 = 32$; write down the 2 and carry the 3; $4 \times 5 = 20 + 3 = 23$. In the third row, $9 \times 0 = 0$; $9 \times 8 = 72$; write down the 2 and carry the 7; $9 \times 5 = 45 + 7 = 52$. Then add.

Here's another set of problems:

```
  905          837
 ×658         ×405
```

Solutions:

```
   905           837
  ×658          ×405
  7240          4185
 4525           000
5430          3348
595490        338985
```

Try to solve this pair of problems:

```
  416          590
 ×807         ×472
```

Solutions:

```
   416           590
  ×807          ×472
  2912          1180
  000          4130
3328          2360
335712        278480
```

5 Four-Digit Multiplication

Now it's on to four-digit multiplication. Let's work out these two problems:

```
 5076         6975
×3245        ×4056
```

Solutions:

$$\begin{array}{r} 5076\\ \times 3245\\ \hline 25380\\ 20304\\ 10152\\ 15228\\ \hline 16471620 \end{array} \qquad \begin{array}{r} 6975\\ \times 4056\\ \hline 41850\\ 34875\\ 0000\\ 27900\\ \hline 28290600 \end{array}$$

If you got these right, you can skip to frame 6 of this chapter. If you got either (or both) of these wrong, here's your chance to redeem yourself:

$$\begin{array}{r} 1904\\ \times 6578\\ \hline \end{array} \qquad \begin{array}{r} 4714\\ \times 5062\\ \hline \end{array}$$

Solutions:

$$\begin{array}{r} 1904\\ \times 6578\\ \hline 15232\\ 13328\\ 9520\\ 11424\\ \hline 12524512 \end{array} \qquad \begin{array}{r} 4714\\ \times 5062\\ \hline 9428\\ 28284\\ 0000\\ 23570\\ \hline 23862268 \end{array}$$

Okay. Now it's time to see how well you can do multiplication with several digits. Do the following problems and check your work. If you were correct on all five, go on to frame 6. If not, go back and redo frames 1–5 and retake Self-Test 1.

SELF-TEST 1

1. $\begin{array}{r} 95\\ \times 32\\ \hline \end{array}$ **2.** $\begin{array}{r} 781\\ \times 450\\ \hline \end{array}$ **3.** $\begin{array}{r} 446\\ \times 108\\ \hline \end{array}$ **4.** $\begin{array}{r} 7509\\ \times 5124\\ \hline \end{array}$ **5.** $\begin{array}{r} 9087\\ \times 1703\\ \hline \end{array}$

ANSWERS

1. $\begin{array}{r} 95\\ \times 32\\ \hline 190\\ 285\\ \hline 3040 \end{array}$ **2.** $\begin{array}{r} 781\\ \times 450\\ \hline 39050\\ 3124\\ \hline 351450 \end{array}$ **3.** $\begin{array}{r} 446\\ \times 108\\ \hline 3568\\ 4460\\ \hline 48168 \end{array}$ **4.** $\begin{array}{r} 7509\\ \times 5124\\ \hline 30036\\ 15018\\ 7509\\ 37545\\ \hline 38476116 \end{array}$ **5.** $\begin{array}{r} 9087\\ \times 1703\\ \hline 27261\\ 636090\\ 9087\\ \hline 15475161 \end{array}$

6 Word Problems with Multiplication

Everyone eventually can multiply one number by another. That's straightforward. But if the same problem is stated in words rather than numbers, it becomes much more confusing to many people. To get over this fear and confusion, let's tackle some word problems.

Before we start, remember that hidden in each of these word problems is one ordinary multiplication problem. All you'll need to do is set up each multiplication problem and solve it. Got that? Then, as the late Jackie Gleason used to say, "And away we go!"

Problem 1:

Bloomingdale's is having a sale on French perfume at $90 a bottle. How much would you have to pay for 12 bottles?

Solution:

```
  $90
  ×12
  180
  90
$1080
```

Problem 2:

An appliance store sold 234 VCRs at $235 each. How much is the store's total sales of VCRs?

Solution:

```
  $235
  ×234
   940
  705
 470
$54990
```

Have you gotten the hang of it? Here's one last one.

Problem 3:

If light travels at the rate of 186,000 miles per second, how far does it travel in two minutes? (Remember: there are 60 seconds in one minute, or 120 seconds in two minutes.)

Solution:

```
   186000  miles
     ×120
  3720000
186000
 22320000  miles
```

Now we'll see how well *you* do at translating words into numbers. You'll know after you've taken Self-Test 2.

SELF-TEST 2

1. A Ford dealer sold 75 used cars at $10,995 each. What were her total sales?
2. If a train travels at an average speed of 70 miles per hour, how far has the train traveled in 14 hours?
3. There are 5,280 feet in a mile. If you walked 18 miles, how many feet did you walk?

ANSWERS

1.
```
 $10995
    ×75
  54975
 76965
$824625
```

2.
```
 70  miles
×14
280
70
980  miles
```

3.
```
 5280  feet
  ×18
42240
5280
95040  feet
```

How did you do? If you got these—that's great. If you didn't, that's not so great. But you're in good company. Most people have some trouble with word problems. We'll have a few more at the end of the next chapter.

4 Division

Division is the reverse of multiplication. For instance, $2 \times 5 = 10$; and $10 \div 5 = 2$. So if you can multiply, you can divide.

1 Short Division

We'll deal with short division first, and we'll get to long division in the second part of this chapter. What's the boundary line between the long and the short? If the number we divide by is a single digit, then we're doing short division; if it's more than one digit, then we're doing long division.

Here is a pair of short division problems.

Problems:

$9\overline{)360}$ $\qquad$ $5\overline{)365}$

Solutions:

$$\begin{array}{r} 40 \\ 9\overline{)360} \end{array} \qquad \begin{array}{r} 73 \\ 5\overline{)36^{1}5} \end{array}$$

If you got both right, go on to the next problem set; if not, read on.

Let's go over the first problem. We know that 9 won't go into 3, but it will go into 36. How many times? More than once? More than two times? More than three times? In fact, 9 goes into 36 exactly 4 times. Remember your multiplication table?

And next, how many times does 9 go into 0? Zero times. So the answer to our problem is 40. Let's check it out. How much is 9×40? It's 360.

In the second division problem, how many times does 5 go into 3? No times. (Technically it goes in three-fifths of one time, but luckily we don't have to worry about this.) How many times does 5 go into 36? More than 5? Yes! More than 6? Yes! More than 7? Yes—a little more than 7.

Here, what we do is write 7 as part of our answer and carry 1 (making the 5 into 15). After all, $5 \times 7 = 35$. Since 5 goes into 15 exactly 3 times, we have our answer: 73. This answer is verified by the multiplication problem $73 \times 5 = 365$.

The check is an important part of division, because by multiplying our answer by the number by which we divided (the divisor), we should end up with the number we divided (the dividend). If not, we must go back and find our mistake.

Here's the next set.

Problems:

8)150   3)191

Solutions:

$$8\overline{)15^{7}0.^{6}0^{4}0} = 18.75 \qquad 3\overline{)19^{1}1.^{2}0^{2}00} = 63.666$$

How did you do? If you got these right, go on to the next problem set; if you didn't, then read the explanation here.

The first problem worked out once we put in a decimal point right after 150 and added a couple of zeros. What gives us the right to do this? Mathematically, every whole number has an unwritten decimal point immediately to its right, followed by any number of zeros. For example, 8 may be written 8.0, or 8.00, or 8.000. The reason we can place a decimal point after a number and add zeros is because doing so doesn't alter the value of that number.

In the first problem, we added a decimal point after 150 and two zeros. By continuing the division, we came up with an answer of 18.75. We need to be very careful about placing our decimal immediately to the right of a number *before* we add zeros. How many zeros do we need to add? Usually just one or two. Beyond that, we will almost always find that we have a repeating answer. This is true in the second problem. When we divide 3 into 191, we get 63, with two left over. No matter how far we carry our division, we'll keep coming up with sixes. So we need to round off our answer. If we were to do so to two decimals, we'd get 63.67. If we rounded to one decimal, we'd have 63.7.

Rounding

Suppose you have $16.25 in your pocket. How much do you have to the nearest dollar? You have $16. If you earn $28,603, how much do you earn to the nearest thousand? You earn $29,000.

If you weigh 117.476 pounds, how much do you weigh to the nearest pound? The answer is 117 pounds. What is your weight to the nearest tenth of a pound? It is 117.5. And your weight to the nearest hundredth of a pound is 117.48.

When we round a number ending in 5, the general practice is to round up (from 115.25 to 115.3) rather than down (from 115.25 to 115.2). However, if you can remember, the soundest practice is to alternate, by rounding upward in the first instance, downward in the second, and so forth.

Let's do another set of problems.

Problems:

$9\overline{)179}$ $\qquad$ $4\overline{)3142}$

Solutions:

$\begin{array}{r} 19.888 \\ 9\overline{)17^{8}9.^{8}0^{8}0^{8}0} \end{array} = 19.89$ $\qquad$ $\begin{array}{r} 785.5 \\ 4\overline{)31^{3}4^{2}2.^{2}0} \end{array}$

If you came up with the right answers to these last two problems, go on to frame 2; if not, try some more problems.

Problems:

$7\overline{)473}$ $\qquad$ $6\overline{)9613}$

Solutions:

$$\begin{array}{r} 67.57 \\ 7\overline{)47^{5}3.^{4}0^{5}0} \end{array} \qquad \begin{array}{r} 1602.166 \\ 6\overline{)9^{3}613.^{1}0^{4}0^{4}0} \end{array}$$

How did you do? Here's one last set.

Problems:

$8\overline{)9077}$ $\qquad$ $5\overline{)8154}$

Solutions:

$$\begin{array}{r} 1134.625 \\ 8\overline{)9^{1}0^{2}7^{3}7.^{5}0^{2}0^{4}0} \end{array} \qquad \begin{array}{r} 1630.8 \\ 5\overline{)8^{3}1^{1}54.^{4}0} \end{array}$$

At this point you'll need to make a judgment. If you're having a tough time, the best thing to do would be to go back to the beginning of this chapter and start again from scratch. You won't be able to do the long division of the next section until you're good at short division.

One of the things we've been stressing is that mathematics is a set of tools that build on one another. You need to know the multiplication table in order to multiply. You can't divide until you learn how to multiply. And you can't do long division until you learn short division.

SELF-TEST 1

1. $6\overline{)450}$ **2.** $7\overline{)497}$ **3.** $4\overline{)183}$ **4.** $5\overline{)286}$ **5.** $9\overline{)790}$

ANSWERS

1. $\begin{array}{r} 75 \\ 6\overline{)45^{3}0} \end{array}$ **2.** $\begin{array}{r} 71 \\ 7\overline{)497} \end{array}$ **3.** $\begin{array}{r} 45.75 \\ 4\overline{)18^{2}3.^{3}0^{2}0} \end{array}$ **4.** $\begin{array}{r} 57.2 \\ 5\overline{)28^{3}6.^{1}0} \end{array}$ **5.** $\begin{array}{r} 87.77 = 87.8 \\ 9\overline{)79^{7}0.^{7}0^{7}0} \end{array}$

2 Long Division

Long division is carried out in two steps: (1) trial and error and (2) multiplication. The process is identical to short division, but it involves a lot more calculation. That's why it's so important to have memorized the multiplication table.

Because these problems are relatively long, we'll take them one at a time and round the answers to the nearest tenth.

Problem:

```
43)795
```

Solution:

```
   18.48 = 18.5
43)795.00
 -43 x xx
  365
 -344
   21 0
  -17 2
    3 80
   -3 44
      36
```

I'd like you to notice a few things. First, we place the decimal point after the 5 and add two zeros. Next, since 43 will go into 79 one time, we multiply 1×43 and place 43 under 79. We subtract 43 from 79, leaving 36. Next, the 5 is brought down (we keep track of this by placing an *X* under the 5). How many times does 43 go into 365? We find, by trial and error, that it goes in eight times.

Now $8 \times 43 = 344$. We subtract 344 from 365 and get 21. The first zero is brought down, and we find that 43 goes into 210 four times. Since $4 \times 43 = 172$, we subtract 172 from 210 and get 38. Next, the second zero is brought down, giving us 380. How many times does 43 go into 380? Eight times: $8 \times 43 = 344$.

That's as far as we go. Since we're interested in carrying our answer to just one decimal, we round off 18.48 to 18.5.

Problem 1:

```
57)2075
```

Solution:

```
     36.4
57)2075.0
  -171 x x
    365
   -342
     23 0
    -22 8
        2
```

How many times does 57 go into 207? It goes in three times, so we write 3 over the 7, multiply 3×57 and get 171. We subtract 171 from 207 and get 36. Then we bring down the next number, 5, which gives us 365. How many times does 57 go into 365? It goes in six times, so we multiply 57×6 and get 342. We subtract 342 from 365 to get 23, and bring down the zero to give us 230. How many times does 57 go into 230? It goes in four times: 4×57=228. Then we subtract 228 from 230 to get 2.

We'll add one more wrinkle and that will be it for long division. We'll divide by three numbers. Try this one on for size.

Problem 2:

```
614)1437
```

Solution:

```
       2.34=2.3
614)1437.00
   -1228 xx
     209 0
    -184 2
      24 80
     -24 56
```

And now, one last problem.

Problem 3:

```
591)83902
```

Solution:

```
      141.96=142.0
591)83902.00
   -591  xx xx
    2480
   -2364
     1162
     -591
      571 0
     -531 9
       39 10
      -35 46
```

Wasn't that a whole lot of fun? Now see what you can do with the problems in Self-Test 2.

SELF-TEST 2

1. $19\overline{)306}$ **2.** $67\overline{)541}$ **3.** $239\overline{)1975}$ **4.** $641\overline{)84330}$ **5.** $785\overline{)75411}$

ANSWERS

1.

```
    16.1
19)306.0
  -19 x x
   116
  -114
     2 0
    -1 9
```

2.

```
     8.07=8.1
67)541.00
  -536 xx
     5 00
    -4 69
```

3.

```
      8.26=8.3
239)1975.00
   -1912 xx
      63 0
     -47 8
      15 20
     -14 34
```

4.

```
     131.56=131.6
641)84330.00
   -641 xx xx
    2023
   -1923
     1000
     -641
      359 0
     -320 5
       38 50
      -38 46
```

5.

```
      96.06=96.1
785)75411.00
   -7065 x xx
     4761
    -4710
       51 00
      -47 10
        3 90
```

3 Word Problems with Division

Here they come again. I hope these won't be traumatic. Just remember: make each word problem into a simple numerical problem.

Problem 1:

On a religious holiday, one-eighth of the 5,000-member student body of James Madison High School stayed home to pray. How many students were absent and how many came to school?

Solution:

```
  6 2 5  stayed home          5000
8)50²0⁴0                     - 625
                              4375 attended
```

Problem 2:

In a school yard there are 264 children. How many soccer teams of 11 players each can be formed?

Solution:

$$
\begin{array}{r}
24 \\
11\overline{)26^{4}4}
\end{array}
$$

Problem 3:

Melissa Jones earned $40,000 last year. How much did she earn per week?

Solution:

$$
\begin{array}{r}
\$769.23 \\
52\overline{)\$40000.00} \\
\underline{-364}^{xx\ xx} \\
360 \\
\underline{-312} \\
480 \\
\underline{-468} \\
120 \\
\underline{-104} \\
160 \\
\underline{-156}
\end{array}
$$

Now try your luck at this Self-Test.

SELF-TEST 3

1. If a rope is 144 inches long, how many yards long is that rope?
2. A government agency had a payroll of $7,500,000 for 864 workers who all earned exactly the same annual pay. How much did each worker earn?
3. Spuds, the neighbor's dog, chews up 16,316 bones a year. How many bones does he go through in a day?

1. $$
\begin{array}{r}
4 \text{ yards} \\
36\overline{)144} \\
\underline{-144}
\end{array}
$$

2.

```
              $8680.555 = $8,680.56
     864)$7500000.000
         −6912xxx xxx
           5880
          −5184
            6960
           −6912
             480 0
            −432 0
              48 00
             −43 20
               4 800
              −4 320
                 480
```

3.

```
                44.7 bones
     365)16316.0
         −1460x x
            1716
           −1460
             256 0
            −255 5
```

5 Fast Multiplication and Division

Did you know that you could multiply many numbers by 10 just by adding a zero? And that you could divide many numbers by 100 by shifting a decimal point two places to the left? When you've completed this chapter you'll be able to multiply and divide in just seconds.

1 Fast Multiplication

One of the laws of arithmetic is that any number multiplied by 1 *is* that number. So the next time someone asks you to multiply 10 by 1, you'll know the answer is 10. And 3×1 is obviously 3.

Next we'll multiply by 10. How much is 16×10? It's 160. What we've done, in effect, is move the decimal one place to the right (from 16.0 to 160.0).

Problems:

Multiply these numbers by 10:

a. 4 b. 37 c. 596

Solutions:

a. $4 \times 10 = 40$ (4.0 to 40.0)

b. $37 \times 10 = 370$ (37.0 to 370.0)

c. $596 \times 10 = 5{,}960$ (596.0 to 5,960.0)

You probably noticed that when we multiplied by 10, what we were really doing was adding a zero to the number being multiplied. That observation is fine as long as we're dealing with whole numbers. But when we deal with decimals, we've got to worry about the decimal point.

How much is $10 \times .9$? The answer is 9.0, or 9. How did we get that? We moved the decimal one place to the right. But what if we had merely added a zero? Then we would have gotten .90, which, mathematically, is equal to .9. So when we're multiplying a decimal by 10, we have to make sure to move the decimal one place to the right. Let's try some problems with decimals.

Problems:

Multiply these numbers by 10:

a. 6.3 b. .4 c. .02

Solutions:

a. 63 b. 4 c. .2

Our next step is to multiply by 100. To do that we add two zeros to whole numbers and move the decimal point two places to the right of the decimals.

Problems:

Multiply each of these numbers by 100:

a. 14 b. 1 c. .7 d. .05

Solutions:

a. 1,400 b. 100 c. 70 d. 5

Let's see what you've learned so far. See if you can complete Self-Test 1 in less than two minutes.

SELF-TEST 1

1. Multiply each of these numbers by 10:
   a. .09 b. .7 c. 86 d. 5.6 e. 102
2. Multiply each of these numbers by 100:
   a. 4 b. .02 c. 17 d. .13 e. .008

ANSWERS

1. a. .9 b. 7 c. 860 d. 56 e. 1,020
2. a. 400 b. 2 c. 1,700 d. 13 e. .8

2 Fast Division

Fast division is the exact reverse of fast multiplication. Instead of moving the decimal place to the right, we move it to the left. And in the case of whole numbers ending with zeros, we subtract zeros.

Are you ready for some problems? Please work these out.

Problems:

Divide each of the following numbers by 10:

a. 800 b. 16 c. .3

Solutions:

a. 80 b. 1.6 c. .03

In each case what we did was move the decimal one place to the left. In the first case, 800.0 became 80.0; in the second, 16 became 1.6; and in the third, .3 became .03.

Now we'll divide by 100. All you'll do here is move the decimal *two* places to the left.

Problems:

Please divide each of these numbers by 100:

a. .6 b. 100 c. 9

Solutions:

a. .006 b. 1 c. .09

I'll bet you're finding these pretty easy. You'll probably be able to complete Self-Test 2 in less than five minutes.

SELF-TEST 2

1. Divide each of these numbers by 10:
   a. .01 b. 4 c. 900 d. .71 e. .3
2. Divide each of these numbers by 100:
   a. 80 b. .14 c. 3.7 d. 916 e. .05

ANSWERS

1. a. .001 b. .4 c. 90 d. .071 e. .03
2. a. .8 b. .0014 c. .037 d. 9.16 e. .0005

3 How Are You Doing So Far?

If the pace we've been setting is too fast for you, then you would do well to turn back to the beginning of chapter 2. Mathematics is a lot like children's toy blocks. If you don't get the bottom blocks down solidly, your building will come tumbling down. Even if this sets you back a few days, in the long run, you'll do much better.

If the pace is too slow for you, then go ahead and try the self-tests in chapter 6. If you do well, then you may skip that chapter and go directly to chapter 7. And if you feel confident, go ahead and try the self-tests in that chapter. But don't skip even a part of any chapter unless you've proven to yourself that you can do the accompanying self-test questions.

6 Multiplying and Dividing with Decimals

Here's something that can't be done very easily on a calculator, even though calculators have floating decimals. Multiplying and dividing with decimals probably leads to more mistakes than any other arithmetic operations. And yet, if we follow just one very simple rule, we can avoid all that grief.

1 Multiplying with Decimals

We'll work out a simple problem and then I'll tell you the rule.

Problem:

$$\begin{array}{r} 1.5 \\ \times 1.3 \\ \hline \end{array}$$

Solution:

$$\begin{array}{r} 1.5 \\ \times 1.3 \\ \hline 45 \\ 15 \\ \hline 1.95 \end{array}$$

This is really a two-part problem. The first part, the multiplication, should not present any difficulties. The second part of the problem is figuring out where to place the decimal point.

Here's the rule we've all been waiting for: Add the number of places to the *right* of the decimals in the numbers being multiplied and place the decimal point that number of places to the *left* in the answer. For instance, 1.5 has one number to the right of the decimal point, and 1.3 also has one number to the right of the decimal point. That gives us two numbers to the right of the decimal points. In our answer, we place our decimal point two places to the left. This will become clearer as we do more problems. Here's another one.

Problem:

```
 2.53
×8.6
```

Solution:

```
  2.53
 ×8.6
  1 518
 20 24
 21.758
```

Here we have .53 and .6 for a total of three places. So we go three places to the left in our answer: 21.758. And now, another example.

Problem:

```
  41.59
 ×76.84
```

Solution:

```
     41.59
    ×76.84
     1 6636
    33 272
   249 54
  2911 3
  3195.7756
```

Are you getting these problems right? If not, redo them. If yes, then you're ready for Self-Test 1.

SELF-TEST 1

1.
```
  90.5
 ×7.3
```
2.
```
  10.764
×45.197
```
3.
```
 9.556
 ×1.03
```

ANSWERS

1.
```
   90.5
   ×7.3
  27 15
  633 5
 660.65
```
2.
```
      10.764
     ×45.197
       75348
      96876
     1 0764
    53 820
   430 56
 486.500508
```
3.
```
   9.556
   ×1.03
   28668
  9 5560
 9.84268
```

2 Dividing with Decimals

Division with decimals is a lot easier than multiplication with decimals. All we need to do is align everything. We'll start off with a very simple problem and work our way into the more complex.

Problem:

```
1.9)3.8
```

Solution:

```
                  2
1.9)3.8 = 19)38.0
            −38
```

All we did was move the decimal one place to the right for both numbers.

Here's another one.

Problem:

```
4.52)15.1
```

Solution:

```
                       3.34 = 3.3
4.52)15.1 = 452)15¹10.0
               − 13 56 x
                  1 54 0
                 −1 35 6
                    18 40
```

By now it should be clear that in order to divide we need to make the divisor, or the number to the left, into a whole number. We do that by moving the decimal point one or more places to the right. And what we do to the divisor, we need to do to the dividend, or the number under the division symbol.

SELF-TEST 2

1. 14.3)13

2. .96)8.2

3. .053)2

ANSWERS

1.
```
                  .909=0.91
14.3)13=143)130.000
           - 128 7xx
             1 300
           - 1 287
```

2.
```
               8.54=8.5
.96)8.2=96)820.00
          -768 xx
            52 0
           -48 0
             4 00
            -3 84
```

3.
```
                37.73=37.7
.053)2=53)2000.00
         - 159x xx
           410
          -371
            39 0
           -37 1
             1 90
            -1 59
               31
```

7 Converting Fractions into Decimals and Decimals into Fractions

Virtually every fraction can be expressed as a decimal, and vice versa. You already knew that 1/4 of a dollar is $.25. How would you express $.50 as a fraction? It's half a dollar, or 1/2 of a dollar. To do a lot of math problems, especially when you're doing algebra, you'll need to be able to convert fractions into decimals and decimals into fractions.

Converting Fractions into Decimals

Converting a fraction into a decimal is a simple problem of division.

Problem 1:

Convert 1/4 into a decimal.

Solution:

$$\frac{1}{4} = 4\overset{.25}{\overline{)1.00}}$$

The big question is why do we divide the 4 into the 1. Once again, we can call upon one of the laws of arithmetic. Whenever we have a fraction, it may be read as follows: divide the top number by the bottom number.

You may raise the objection that you can't divide 4 into 1. While it's true that 4 doesn't go into 1 completely, it does go in one-quarter of the way. For example, if a 400-foot train went through a 100-foot tunnel, only one-quarter of the train would be in the tunnel. So we could say that 400 goes into 100 one-quarter (or 100/400) of the way.

When you divide a large number into a smaller one, your quotient, or answer, will be less than 1.

We'll try another.

Problem 2:

Convert 2/5 into a decimal.

Solution:

$$\frac{2}{5} = 5\overline{)2.0}^{\;.4}$$

We'll do one last one.

Problem 3:

Convert 7/10 into a decimal.

Solution:

$$\frac{7}{10} = 10\overline{)7.0}^{\;.7}$$

The problems in Self-Test 1 are like the ones you've just completed.

SELF-TEST 1

1. Convert 3/4 into a decimal.
2. Convert 1/6 into a decimal.
3. Convert 3/8 into a decimal.
4. Convert 1/12 into a decimal.

ANSWERS

1. $\frac{3}{4} = 4\overline{)3.00}^{\;.75}$
2. $\frac{1}{6} = 6\overline{)1.0000}^{\;.1666 = .167}$
3. $\frac{3}{8} = 8\overline{)3.000}^{\;.375}$
4. $\frac{1}{12} = 12\overline{)1.0000}^{\;.0833 = .083}$

2 Converting Decimals into Fractions

Decimals can easily be converted into fractions. For example, .3 becomes 3/10. We did this by moving the decimal point one place to the right, which multiplied .3 by 10. Then we put 3 over 10, which divided 3 by 10.

Think of .3 as a fraction:

$$\frac{.3}{1}$$

Any number may be written over 1 because it does not change its value—1 divided into any number *is* that number. Why did we write .3 over 1? So that we could make it into a fraction. Then we eliminate the decimal point by multiplying .3 by 10. Invoking the famous law of arithmetic—what you do to the top of a fraction, you must do to the bottom—we multiply the bottom by 10:

$$\frac{.3 \times 10 = 3}{1 \times 10 = 10}$$

Want to try one on your own?

Problem 1:

Convert .57 into a fraction.

Solution:

$$.57 = \frac{.57}{1} = \frac{.57 \times 100}{1 \times 100} = \frac{57}{100}$$

Problem 2:

Convert the following three-place decimal into a fraction: .833.

Solution:

$$.833 = \frac{.833}{1} = \frac{.833 \times 1{,}000}{1 \times 1{,}000} = \frac{833}{1{,}000}$$

Now we can step back and generalize. When we're converting a decimal into a fraction, if that decimal has just one place (e.g., .3), then we remove the decimal point and divide by 10. If that decimal has two places (e.g., .57), then we remove the decimal point and divide by 100. And if that decimal has three places (e.g., .833), we remove the decimal point and divide by 1,000. In general, then, for every place we move the decimal point to the right (to remove it), we divide by a 1 followed by the same number of zeros as we moved the decimal point to the right.

SELF-TEST 2

1. Convert .13 into a fraction.
2. Convert .783 into a fraction.
3. Convert .9 into a fraction.
4. Convert .761 into a fraction.

ANSWERS

1. $.13 = \frac{.13}{1} = \frac{.13 \times 100}{1 \times 100} = \frac{13}{100}$
2. $.783 = \frac{.783}{1} = \frac{.783 \times 1{,}000}{1 \times 1{,}000} = \frac{783}{1{,}000}$
3. $.9 = \frac{.9}{1} = \frac{.9 \times 10}{1 \times 10} = \frac{9}{10}$
4. $.761 = \frac{.761}{1} = \frac{.761 \times 1{,}000}{1 \times 1{,}000} = \frac{761}{1{,}000}$

3 Reducing Fractions to Their Lowest Denominators

If you look back at all the problems we had in frame 2 and in Self-Test 2, you were asked to convert odd-numbered decimals into fractions: .3, .57, .833, .13, .783, .9, and .761. You'll see I was deliberately trying to avoid even numbers. Why? Because even numbers would leave us with fractions that would have to be reduced.

Mathematicians, unlike normal citizens, cannot leave unreduced fractions just sitting there. In fact, no mathematician can go to bed at night until his or her fractions have been reduced to their lowest possible denominators (or bottom numbers). Take the fraction 4/10. That might leave *you* satisfied, but I can't tell you how much the sight of a 4 over a 10 would frustrate a mathematician, who would immediately reduce it to 2 /5.

Now I know that you might not lose much sleep over 4/10, but think of all the mathematicians you'd keep up all night by not reducing all of your fractions to their lowest denominators. And not just fractions with even numbers on top and bottom—all fractions that can be reduced.

Problem 1:

Reduce 20/100 to its lowest denominator.

Solution:

$$\frac{20}{100} = \frac{2}{10} = \frac{1}{5}$$

Incidentally, you may have noticed that we invoke that all-important law of arithmetic—what we do to the top number (numerator) of a fraction, we must do to the bottom number (denominator). In this case, we divided the top by 10 and the bottom by 10 (by cancelling the zeros), and then we divided the top and bottom by 2. Of course, if you immediately recognized that 20 went into 100 five times, you could have simply divided top and bottom by 20.

Problem 2:

Reduce 25/500 to its lowest denominator.

Solution:

$$\frac{25}{500}=\frac{1}{20}$$

That's all there is to it. Now we'll combine two operations—converting decimals into fractions and then reducing them to their lowest denominators. Are you ready? Here comes the first one.

Problem 3:

Convert .45 into a fraction.

Solution:

$$.45=\frac{.45}{1}=\frac{.45\times 100}{1\times 100}=\frac{45}{100}=\frac{9}{20}$$

Here's one more.

Problem 4:

Convert .68 into a fraction.

Solution:

$$.68=\frac{.68}{1}=\frac{.68\times 100}{1\times 100}=\frac{68}{100}=\frac{34}{50}=\frac{17}{25}$$

Now see how you do on Self-Test 3.

SELF-TEST 3

Convert each of these decimals into fractions and reduce them to their lowest denominators.

1. .75 **2.** .8 **3.** .65 **4.** .245

1. $.75=\frac{.75}{1}=\frac{.75\times 100}{1\times 100}=\frac{75}{100}=\frac{3}{4}$

2. $.8=\frac{.8}{1}=\frac{.8\times 10}{1\times 10}=\frac{8}{10}=\frac{4}{5}$

3. $.65=\frac{.65}{1}=\frac{.65\times 100}{1\times 100}=\frac{65}{100}=\frac{13}{20}$

4. $.245=\frac{.245}{1}=\frac{.245\times 1,000}{1\times 1,000}=\frac{245}{1,000}=\frac{49}{200}$

8 Adding, Subtracting, Multiplying, and Dividing Fractions

When you're having a large party catered, do you order new or used food? It's pretty hard to order used food, but if you could, it would certainly save you a lot of money. Imagine you needed a total of two cakes and you told the caterer to give you parts of a few. You might end up with 1/2 of a pineapple upside-down cake, 1/2 of a Black Forest cake, 1/2 of a chocolate layer cake, and 1/2 of a cheesecake. Add up these half cakes and see what you get.

You would get $1/2 + 1/2 + 1/2 + 1/2 = 2$ cakes. You've just done your first problem adding fractions. For you, it was a piece of cake—or, actually, four pieces of cake.

1 Adding Fractions

Fractions, like whole numbers, can be added, subtracted, multiplied, and divided. First we'll add them.

Problem 1:

Add 1/2 and 1/2.

Solution:

$$\frac{1}{2}+\frac{1}{2}=\frac{2}{2}=1$$

That was an easy one. Here's something a bit harder.

Problem 2:

Add 1/6 and 2/6.

Solution:

$$\frac{1}{6}+\frac{2}{6}=\frac{3}{6}=\frac{1}{2}$$

Let's run the videotape for our instant replays. Everybody knows that $1/2 + 1/2 = 1$. But what did we really do in terms of numerators (top numbers) and denominators (bottom numbers)? We added the numerators and wrote their sum over a common denominator.

$$\frac{1+1}{2}=\frac{2}{2}=1$$

We did the same thing when we added $1/6 + 2/6$. We added the numerators over the common denominator:

$$\frac{1+2}{6}=\frac{3}{6}=\frac{1}{2}$$

Notice that we've reduced this fraction to its lowest terms. Why? Mainly so that it is in its most recognizable form. For example, maybe *you* know that $176/352 = 1/2$, but not everyone else does. If you reduce your fractions to their lowest terms, then anyone looking at your answers would know exactly how much they were. And as we noted in frame 3 of the last chapter, reducing a fraction to its lowest denominator keeps our mathematicians very happy.

When we add two or more fractions, we need to find the lowest common denominator. If you were to add 1/5 and 1/3, you would find that the lowest common denominator is 15. Before we can add 1/5 and 1/3, we need to express each with 15 as its denominator:

$$\frac{1}{5}=\frac{1\times 3}{5\times 3}=\frac{3}{15}$$

$$\frac{1}{3}=\frac{1\times 5}{3\times 5}=\frac{5}{15}$$

From here on out it's easy. Just add:

$$\frac{3}{15}+\frac{5}{15}=\frac{8}{15}$$

Problem 3:

Add 2/3 and 1/6.

Solution:

$$\frac{2}{3}+\frac{1}{6}=\frac{2\times 2}{3\times 2}+\frac{1}{6}=\frac{4}{6}+\frac{1}{6}=\frac{5}{6}$$

Do you follow what we did here? We wanted to give both fractions a common denominator so we could add them. To convert 2/3 into 4/6, we multiplied the top and the bottom of the fraction by 2. Remember the law of arithmetic that allows this? What you do to the top of a fraction, you must do to the bottom. So we multiplied the top by 2 and the bottom by 2.

Here's another one.

Problem 4:

Add 1/3 and 2/5.

Solution:

$$\frac{1}{3}+\frac{2}{5}=\frac{1\times 5}{3\times 5}+\frac{2\times 3}{5\times 3}=\frac{5}{15}+\frac{6}{15}=\frac{11}{15}$$

To repeat, we want the lowest common denominator so we don't have to reduce our answer (or, if it can be further reduced, we'd be able to minimize that reduction). And we want a common denominator so we'll be adding units of the same thing. Just as you can't add apples and oranges, you can't add thirds and quarters without finding their (lowest) common denominator, which happens to be 12.

And now for one more problem.

Problem 5:

Add 3/4, 1/3, and 1/6.

Solution:

First find the lowest common denominator. Clearly, it is 12: 4 goes into 12, 3 goes into 12, and 6 goes into 12.

$$\frac{3}{4}+\frac{1}{3}+\frac{1}{6}=\frac{3\times 3}{4\times 3}+\frac{1\times 4}{3\times 4}+\frac{1\times 2}{6\times 2}$$
$$=\frac{9}{12}+\frac{4}{12}+\frac{2}{12}=\frac{15}{12}=1\frac{3}{12}=1\frac{1}{4}$$

As you've noticed, 15/12 is greater than 1. When we convert it to 1 3/12, we are converting a fraction into a mixed number (a whole number and a fraction). The fraction 15/12 tells us to divide 12 into 15, which gives us 1 3/12. We'll encounter more mixed numbers in subsequent frames of this chapter and later chapters.

Are you getting the hang of it? Only you know for sure. If you've gotten the last two problems right, go on to Self-Test 1. Otherwise, please return to frame 1.

SELF-TEST 1

1. $\frac{1}{4}+\frac{1}{5}$
2. $\frac{3}{10}+\frac{2}{5}$
3. $\frac{1}{2}+\frac{1}{3}+\frac{3}{4}$
4. $\frac{1}{6}+\frac{2}{3}+\frac{4}{9}$

ANSWERS

1. $\frac{1}{4}+\frac{1}{5}=\frac{1\times 5}{4\times 5}+\frac{1\times 4}{5\times 4}=\frac{5}{20}+\frac{4}{20}=\frac{9}{20}$
2. $\frac{3}{10}+\frac{2}{5}=\frac{3}{10}+\frac{2\times 2}{5\times 2}=\frac{3}{10}+\frac{4}{10}=\frac{7}{10}$
3. $\frac{1}{2}+\frac{1}{3}+\frac{3}{4}=\frac{1\times 6}{2\times 6}+\frac{1\times 4}{3\times 4}+\frac{3\times 3}{4\times 3}=\frac{6}{12}+\frac{4}{12}+\frac{9}{12}=\frac{19}{12}=1\frac{7}{12}$
4. $\frac{1}{6}+\frac{2}{3}+\frac{4}{9}=\frac{1\times 3}{6\times 3}+\frac{2\times 6}{3\times 6}+\frac{4\times 2}{9\times 2}=\frac{3}{18}+\frac{12}{18}+\frac{8}{18}=\frac{23}{18}=1\frac{5}{18}$

2 Subtracting Fractions

There's absolutely nothing new here except for a change of sign:

Problem 1:

$$\frac{5}{6}-\frac{1}{6}$$

Solution:

$$\frac{5}{6}-\frac{1}{6}=\frac{4}{6}=\frac{2}{3}$$

Problem 2:

$$\frac{1}{3}-\frac{1}{4}$$

Solution:

$$\frac{1}{3}-\frac{1}{4}=\frac{1\times4}{3\times4}-\frac{1\times3}{4\times3}=\frac{4}{12}-\frac{3}{12}=\frac{1}{12}$$

Problem 3:

$$\frac{3}{4}-\frac{2}{5}$$

Solution:

$$\frac{3}{4}-\frac{2}{5}=\frac{3\times5}{4\times5}-\frac{2\times4}{5\times4}=\frac{15}{20}-\frac{8}{20}=\frac{7}{20}$$

SELF-TEST 2

1. $\frac{3}{5}-\frac{1}{3}$
2. $\frac{7}{8}-\frac{3}{4}$
3. $\frac{2}{3}-\frac{2}{5}$
4. $\frac{1}{2}-\frac{1}{8}$

ANSWERS

1. $\frac{3}{5}-\frac{1}{3}=\frac{3\times3}{5\times3}-\frac{1\times5}{3\times5}=\frac{9}{15}-\frac{5}{15}=\frac{4}{15}$
2. $\frac{7}{8}-\frac{3}{4}=\frac{7}{8}-\frac{3\times2}{4\times2}=\frac{7}{8}-\frac{6}{8}=\frac{1}{8}$
3. $\frac{2}{3}-\frac{2}{5}=\frac{2\times5}{3\times5}-\frac{2\times3}{5\times3}=\frac{10}{15}-\frac{6}{15}=\frac{4}{15}$
4. $\frac{1}{2}-\frac{1}{8}=\frac{1\times4}{2\times4}-\frac{1}{8}=\frac{4}{8}-\frac{1}{8}=\frac{3}{8}$

3 Multiplying by Fractions

How do you find a fraction of a fraction? This is a straightforward multiplication problem. So let's set it up.

Problem 1:

How much is one-eighth of one-quarter?

Solution:

Write down one-eighth as a fraction. Then write down one-quarter. Your fractions should look like this:

$$\frac{1}{8}, \frac{1}{4}$$

The final step is to multiply them:

$$\frac{1}{8}\times\frac{1}{4}=\frac{1\times1}{8\times4}=\frac{1}{32}$$

One nice thing about multiplying fractions is that it's not necessary to figure out a common denominator, because you'll find it automatically. But is the 32 in the previous problem the lowest common denominator? It is, in this case. But you'll find that when you multiply fractions, you can often reduce your result to a lower denominator.

Here's another one.

Problem 2:

How much is one-third of one-seventh?

Solution:

$$\frac{1}{3}\times\frac{1}{7}=\frac{1\times1}{3\times7}=\frac{1}{21}$$

Now we'll get fancy.

Problem 3:

How much is one-half of two and a half? Hint: Set up both as fractions. Think of two and a half in terms of halves.

Solution:

$$\frac{1}{2}\times2\frac{1}{2}=\frac{1}{2}\times\frac{5}{2}=\frac{5}{4}=1\frac{1}{4}$$

Two and a half is a mixed number, which doesn't lend itself well to multiplication. So we convert it into a fraction—there are four halves in two, plus the one half, giving us five halves, or 5/2.

Are you getting the hang of it? I hope so. Try this one.

Problem 4:

Find one-third of seven.

Solution:

$$\frac{1}{3}\times\frac{7}{1}=\frac{7}{3}=2\frac{1}{3}$$

The trick here is to write the number 7 in fraction form. Any number divided by 1 *is* that number: 7 divided by 1 is 7. So we are allowed to write 7 as 7/1. We put the 7 in fractional form so we can multiply it by 1/3.

Problem 5:

Here's a chance (if necessary) to redeem yourself. Find two-fifths of nine.

Solution:

$$\frac{2}{5} \times \frac{9}{1} = \frac{18}{5} = 3\frac{3}{5}$$

Ready for another Self-Test? All right, then, here it comes.

SELF-TEST 3

1. Find two-thirds of one-sixth.
2. Find one-half of three and a half.
3. Find three-fifths of six.
4. Find three-quarters of twelve.

ANSWERS

1. $\frac{2}{3} \times \frac{1}{6} = \frac{2}{18} = \frac{1}{9}$ 2. $\frac{1}{2} \times \frac{7}{2} = \frac{7}{4} = 1\frac{3}{4}$ 3. $\frac{3}{5} \times \frac{6}{1} = \frac{18}{5} = 3\frac{3}{5}$ 4. $\frac{3}{4} \times \frac{12}{1} = \frac{36}{4} = 9$

4 Dividing by Fractions

Let's get right into it. How much is one-half divided by one-fourth? Don't panic! The trick to doing this is to convert it into a multiplication problem. Multiply one-half by the reciprocal of one-fourth. The reciprocal of a fraction is found by turning the fraction upside down. So 1/4 becomes 4/1.

Problem 1:

What is one-half divided by one-fourth?

Solution:

$$\frac{1}{2} \times \frac{4}{1} = \frac{4}{2} = 2$$

What gives us the right to convert a division problem into a multiplication problem just by converting one term into its reciprocal? This process can be explained by converting these fractions into decimals. One-half is .50, and one-quarter is .25. How many times does .25 go into .50? Two times. This is the same answer as the one we got when we multiplied 1/2 by 4/1.

We'll try another problem and then show how it could have been done by division of decimals.

Problem 2:

How much is one-fifth divided by one-twentieth?

Solution:

$$\frac{1}{5} \times \frac{20}{1} = \frac{20}{5} = 4$$

Problem 3:

How much is the decimal of one-fifth divided by the decimal of one-twentieth?

Solution:

$$\frac{1}{5} = \begin{array}{r} .20 \\ 5\overline{)1.00} \end{array} \qquad \frac{1}{20} = \begin{array}{r} .05 \\ 20\overline{)1.00} \end{array} \qquad \begin{array}{r} \\ .05\overline{).20} \end{array} = \begin{array}{r} 4 \\ 5\overline{)20} \end{array}$$

Just remember to flip the fraction you're dividing by and you've set up your multiplication problem. Now let's see how you do on Self-Test 4.

SELF-TEST 4

1. Find one-sixth divided by one-fourth.

2. Find two-fifths divided by two-thirds.

3. Find three-quarters divided by one-eighth.

4. Find one-third divided by two-fifths.

1. $\frac{1}{6} \times \frac{4}{1} = \frac{4}{6} = \frac{2}{3}$ **2.** $\frac{2}{5} \times \frac{3}{2} = \frac{6}{10} = \frac{3}{5}$ **3.** $\frac{3}{4} \times \frac{8}{1} = \frac{24}{4} = 6$ **4.** $\frac{1}{3} \times \frac{5}{2} = \frac{5}{6}$

5 Cancelling Out

When was the last time you cancelled out? No, this is not a reference to your untimely demise, nor even to your physical or mental state. It's an arithmetic term.

Wouldn't you agree that it's a lot easier to multiply and divide relatively small numbers rather than large numbers? I thought so. Well, cancelling out is a way of reducing large numbers to small ones.

In this problem, we'll do just that:

(long way) $\frac{15}{34} \times \frac{17}{5} = \frac{255}{170} = 1\frac{1}{2}$

(easy way) $\frac{\overset{3}{\cancel{15}}}{\underset{2}{\cancel{34}}} \times \frac{\overset{1}{\cancel{17}}}{\underset{1}{\cancel{5}}} = \frac{3}{2} = 1\frac{1}{2}$

Isn't 3/2 a lot nicer than 255/170? We divided: 5 by 5 to get 1; 15 by 5 to get 3; 17 by 17 to get 1; and 34 by 17 to get 2.

Please simplify and solve this problem:

$\frac{5}{9} \times \frac{8}{25} = \frac{\overset{1}{\cancel{5}}}{9} \times \frac{8}{\underset{5}{\cancel{25}}} = \frac{8}{45}$

Exactly what did we do here? We crossed out the 5 and wrote 1, and we crossed out the 25 and wrote 5. So 5/9 × 8/25 became 1/9 × 8/5. What we really did was divide 5 by 5 and 25 by 5, cancelling out the 5s.

What gives us the right to do this cancelling out? Remember the arithmetic rule that what you do to the top of a fraction, you must do to the bottom? You may claim that 5/9 and 8/25 are separate fractions, but I would be forced to disagree with you since they could be stated as:

$\frac{5 \times 8}{9 \times 25}$

Problem:

$\frac{7}{18} \times \frac{9}{21}$

Solution:

$$\frac{\overset{1}{\cancel{7}}}{\underset{2}{\cancel{18}}} \times \frac{\overset{1}{\cancel{9}}}{\underset{3}{\cancel{21}}} = \frac{1}{6}$$

One more Self-Test and we're out of here.

SELF-TEST 5

1. $\frac{45}{50} \times \frac{3}{9}$
2. $\frac{36}{4} \times \frac{16}{6}$
3. $\frac{12}{20} \times \frac{60}{24}$

ANSWERS

1. $\frac{\overset{1}{\cancel{45}^{5}}}{\underset{10}{\cancel{50}}} \times \frac{3}{\underset{1}{\cancel{9}}} = \frac{3}{10}$
2. $\frac{\overset{6}{\cancel{36}}}{\underset{1}{\cancel{4}}} \times \frac{\overset{4}{\cancel{16}}}{\underset{1}{\cancel{6}}} = \frac{24}{1} = 24$
3. $\frac{\overset{1}{\cancel{12}}}{\underset{1}{\cancel{20}}} \times \frac{\overset{3}{\cancel{60}}}{\underset{2}{\cancel{24}}} = \frac{3}{2} = 1\frac{1}{2}$

9 Advanced Word Problems

Oh no! Not more word problems! Sorry, but I'll tell you what. Let's make a deal. Try to do the first couple of problems. If you can do them, go on to the next couple. And if those work out, keep going.

And what if you don't get the first couple right? Well, it's really up to you. I can't twist your arm. I mean, if you absolutely hate word problems, then just skip the rest of the chapter and try your hand at chapter 10. But first, at least try to do the first two problems of this chapter.

1 Words and Fractions

Problem 1:

How much is one-quarter of one-third?

Solution:

$$\frac{1}{3} \times \frac{1}{4} = \frac{1}{12}$$

Problem 2:

Last year, Jason was 4 feet 8 1/8 inches tall. Now he is 5 feet 1 1/4 inches tall. How much did he grow over the last year? (Hint: Convert feet into inches.)

Solution:

$$5'1\frac{1}{4}'' - 4'8\frac{1}{8}'' = 61\frac{1}{4}'' - 56\frac{1}{8}'' = 61\frac{1'' \times 2}{4 \times 2} - 56\frac{1}{8}''$$

$$= 61\frac{2}{8}'' - 56\frac{1}{8}'' = 5\frac{1}{8}''$$

Problem 3:

At the weight loss center last week, Max lost 3 1/2 pounds, Karen lost 2 7/8 pounds, and Sharon lost 1 3/4 pounds. How much weight did the three of them lose all together?

Solution:

$$3\frac{1}{2} + 2\frac{7}{8} + 1\frac{3}{4} = \frac{7}{2} + \frac{23}{8} + \frac{7}{4} = \frac{7 \times 4}{2 \times 4} + \frac{23}{8} + \frac{7 \times 2}{4 \times 2}$$

$$= \frac{28}{8} + \frac{23}{8} + \frac{14}{8} = \frac{65}{8} = 8\frac{1}{8} \text{ pounds}$$

An alternate method would be to add the whole numbers and fractions separately:

$$3 + 2 + 1 = 6$$

$$\frac{1}{2} + \frac{7}{8} + \frac{3}{4} = \frac{4}{8} + \frac{7}{8} + \frac{6}{8} = \frac{17}{8} = 2\frac{1}{8}$$

$$6 + 2\frac{1}{8} = 8\frac{1}{8}$$

Problem 4:

John bought two sets of weights totaling 315 1/2 pounds. If the first set weighed 132 3/4 pounds, how much did the second set weigh?

Solution:

$$315\frac{1}{2} - 132\frac{3}{4} = 315\frac{1}{2}$$

$$\underline{-132\frac{3}{4}}$$

How can we subtract 3/4 from 1/2? After all, 3/4 is larger than 1/2. The trick is to borrow 1 from 315. That 1 may be expressed as 4/4. Adding 4/4 to 1/2 (or 2/4) gives us 6/4. Here it is, step by step:

$$\begin{array}{r} 315\frac{1}{2} \\ -132\frac{3}{4} \\ \hline \end{array} = \begin{array}{r} 315\frac{2}{4} \\ -132\frac{3}{4} \\ \hline \end{array} = \begin{array}{r} 314\frac{6}{4} \\ -132\frac{3}{4} \\ \hline 182\frac{3}{4} \end{array}$$

Problem 5:

Elizabeth bought 12 1/3 yards of material to make dresses. She used 3 1/2 yards on the first dress and 3 7/8 on the second. How much material was left over?

Solution:

$$12\frac{1}{3}-\left(3\frac{1}{2}+3\frac{7}{8}\right)=12\frac{1}{3}-\left(3\frac{1\times 4}{2\times 4}+3\frac{7}{8}\right)=12\frac{1}{3}-\left(3\frac{4}{8}+3\frac{7}{8}\right)$$
$$=12\frac{1}{3}-\left(6+\frac{11}{8}\right)=12\frac{1}{3}-\left(6+1\frac{3}{8}\right)$$
$$=12\frac{1}{3}-7\frac{3}{8}=\frac{37}{3}-\frac{59}{8}$$
$$=\frac{37\times 8}{3\times 8}-\frac{59\times 3}{8\times 3}=\frac{296}{24}-\frac{177}{24}=\frac{119}{24}=4\frac{23}{24}$$

Problem 6:

How many strips of wood 5/8 of an inch wide can be sawed off a 5-foot piece of wood?

Solution:

$$60" \div \frac{5"}{8}=\frac{\overset{12}{\cancel{60}}}{1}\times\frac{8}{\underset{1}{\cancel{5}}}=96$$

2 Word Problems with Money

Problem 1:

How much would it cost to buy 4 2/3 yards of cloth at $11 per yard?

Solution:

$$4\frac{2}{3}\times \$11=\frac{14}{3}\times\frac{\$11}{1}=\frac{\$154}{3}=\$51.33$$

Problem 2:

If silver wire were sold for 40 cents an inch, how much would it cost to buy 2 yards of wire?

Solution:

Remember to convert: 2 yards = 6 feet = 72 inches.
72 × \$.40 = \$28.80

Problem 3:

Farmer Jones bought 7 1/2 bales of hay for \$30. How much did 1 bale cost?

Solution:

What we really have here is a multiplication problem masquerading as a division problem. We could divide \$30 by 7 1/2 and get our cost per bale:

$$\frac{\$30}{7\frac{1}{2}}=\frac{\$30}{7.5}=7.5\overline{)\$30}=75\overline{)\$300}\quad \begin{array}{r}4\\ \$300\\ -\underline{300}\end{array}$$

But here's an easier way:

$$\$30\div 7\frac{1}{2}=\frac{\$30}{1}\div\frac{15}{2}=\frac{\$30}{1}\times\frac{2}{15}=\frac{\overset{2}{\cancel{\$30}}}{1}\times\frac{2}{\underset{1}{\cancel{15}}}=\$4$$

If you'd like to review this type of operation, you'll find it in frames 4 and 5 of chapter 8. It should be emphasized that there's no one correct method to solve most mathematical problems. Like accountants, mathematicians are most concerned with the bottom line.

Let's see how you do on this self-test.

SELF-TEST 1

1. How much is two-sevenths of one-third?
2. Gillian and Amanda went on diets. Together they lost 20 3/4 pounds. If Amanda lost 12 1/8 pounds, how much did Gillian lose?

3. If you walked 23 miles at an average speed of 13 1/4 minutes per mile, how long did it take you to walk the entire distance?

4. If gold were selling at $450 an ounce, how much gold could you buy for $7,875?

ANSWERS

1. $\frac{1}{3} \times \frac{2}{7} = \frac{2}{21}$

2. $20\frac{3}{4} - 12\frac{1}{8} = 20\frac{3 \times 2}{4 \times 2} - 12\frac{1}{8} = 20\frac{6}{8} - 12\frac{1}{8} = 8\frac{5}{8}$ pounds

3. $\frac{23}{1} \times \frac{53}{4} = \frac{1,219}{4} = 304\frac{3}{4}$ minutes, or 5 hours, 4 minutes, and 45 seconds

4. $450\overline{)7875} = 45\overline{)787.5}$ = 17.5 ounces

$$\begin{array}{r} 17.5 \\ 45\overline{)787.5} \\ -45 \\ \hline 337 \\ -315 \\ \hline 22\,5 \\ -22\,5 \\ \hline \end{array}$$

10 Percentages

Nothing in life, not even Ivory Soap, seems to be 100 percent pure. Ninety-nine and forty-four one-hundredths percent pure is just fine, thank you. The only trouble is, exactly *how* do you calculate percentages? Funny you should ask.

1 Fractions, Decimals, and Percentages

Have you ever wondered what a percent is? Percentages can be expressed in fraction and decimal form—for instance, 1% is 1/100 or one-hundredth. How much is 4%? Right—it's 4/100. And how much is 43%? It's 43/100. How much is 43/100 in decimal form? The answer is .43. To summarize, 43% = 43/100 = .43.

Usually, a percentage is expressed by using a percent sign, so we can say that 76% = 76/100 = .76.

To convert a decimal to a percentage, move the decimal point two places to the right and add a percentage sign. For example, convert .23 to a percentage. It's 23%. Convert .06 to a percent. It's 6%. Change .375 to a percent. The answer is 37.5%. (See the box on p. 57, "Percentages as Numbers.")

You probably noticed that all of the fractions so far had 100 as the bottom number. But we won't always be lucky enough to have 100 on the bottom.

Problem 1:

Convert 3/10 into a percent.

Solution:

$$\frac{3 \times 10}{10 \times 10} = \frac{30}{100} = 30\%$$

Percentages as Numbers

Let's take a closer look at the relationship between decimals, fractions, and percentages. We've seen that .01 and 1/100 both equal 1%. How about .10 and 10/100? Obviously, they're equal to 10%.

How much are 1.00 and 100/100 equal to? They must be equal to 100%. In other words, 100% is numerically equal to the number one. Therefore, 1% = 1/100; 10% = 10/100; and 100% = 100/100, or 1.

Problem 2:

Change 1/5 into a percent.

Solution:

$$\frac{1 \times 20}{5 \times 20} = \frac{20}{100} = 20\%$$

Though these are not so hard, there are some numbers that are more difficult to convert.

Problem 3:

Convert 7/8 into a percent.

Solution:

$$\begin{array}{r} .8\ 7\ 5 = 87.5\% \\ 8\,\overline{)7.0^{6}0^{4}0} \end{array}$$

Ready for another?

Problem 4:

Convert 1/6 into a percent.

Solution:

$$6\overline{)1.0^40^40^40} = .1\,6\,6\,6 = 16.7\%$$

Problem 5:

Change 4/3 into a percent.

Solution:

$$3\overline{)4.^10^10^10^10} = 1.\,3\,3\,3\,3 = 133.3\%$$

For a practical example of these conversions see the following box on how to calculate Social Security taxes.

How Your Social Security Tax Is Calculated

The federal government obtains funding for Social Security by collecting 6.2% of your paycheck from you and another 6.2% from your employer. If your salary is $300 a week, how much do *you* pay in Social Security tax?

Solution:

$$\begin{array}{r} \$300 \\ \underline{\times .062} \end{array}$$

or, to make it a little easier,

$$\begin{array}{r} .062 \\ \underline{\times \$300} \\ \$18.600 \end{array}$$

Every week, then, the government gets $18.60 in Social Security tax from your paycheck. And it collects another $18.60 from your employer.

SELF-TEST 1

Convert these fractions into percents:

1. 15/100 **2.** 36/100 **3.** 122/100

Convert these decimals into percents:

4. .75 **5.** 1.66 **6.** .05

Convert these fractions into percents:

7. 3/20 **8.** 4/7 **9.** 3/8

ANSWERS

1. 15% **2.** 36% **3.** 122% **4.** 75% **5.** 166% **6.** 5%

7. $\frac{3\times5}{20\times5}=\frac{15}{100}=15\%$ **8.** $7\overline{)4.0^50^10^30}$ = .5 7 1 4 = 57.1% **9.** $8\overline{)3.0^60^40}$ = .3 7 5 = 37.5%

2 Percentage Changes

Did you know that perhaps nine out of ten college graduates can't work out percentage changes? But don't take *my* word for it. Next June head over to the closest college and see how well the graduates do on Self-Test 2. But we'll let you cheat a little by reading this frame.

Problem 1:

If you were earning \$400 a week and received a \$10 raise, by what percent was your salary increased?

Solution:

$$\frac{\text{change}}{\text{original number}}=\frac{\$10}{\$400}=\frac{1}{40}=2.5\%$$

Problem 2:

If your IQ went from 85 to 140 after reading this book, by what percent did your IQ change?

Solution:

$$\frac{\text{change}}{\text{original number}}=\frac{55}{85}=\frac{11}{17}=64.7\%$$

```
      .647
17 )11.000
  - 10 2xx
       80
     - 68
      120
    - 119
```

Problem 3:

The Atlanta Braves had a better year in 2006 than they had in 2005. In 2006 they won 70 games; in 2005 they won only 56. By what percentage did their wins increase?

Solution:

$$\frac{14}{56} = \frac{2}{8} = \frac{1}{4} = .25 = 25\%$$

If you got this right, go on to Self-Test 2. If you didn't, then you may be wondering where the 14 came from. If the Braves won 70 games in 2006 and 56 in 2005, then the change is 14. If you feel you need further review, go back to the beginning of frame 2.

SELF-TEST 2

1. If you went on a diet and lost 35 pounds, by what percentage did your weight decline if you now weigh 100 pounds?

2. If you grew from 5 feet 3 inches to 5 feet 7 inches, by what percentage did your height increase?

3. If your pay was cut from $55,000 to $52,000, by what percentage did your pay decline?

ANSWERS

1. $\frac{35}{135} = \frac{7}{27}$

```
      .259 = 25.9%
27 )7.000
   -5 4xx
     1 60
    -1 35
      250
     -243
        7
```

2. 5 feet 3 inches = 63 inches; 5 feet 7 inches = 67 inches

$\frac{4}{63}$

```
      .0634 = 6.3%
63 )4.0000
   -3 78xx
      220
     -189
      310
     -252
```

3. $\frac{\$3,000}{\$55,000}$

```
      .0545 = 5.5%
55 )3.0000
   -2 75xx
      250
     -220
      300
     -275
```

3 Fast Percentage Changes

Pick a number, any number. Now triple it. By what percentage did your number increase? Take your time. Use this space to calculate the percentage:

What did you get? Three hundred percent? No, that's incorrect.

As an example, let's use 100. Now let's triple it. We have 300. How much is the percentage increase when you go from 100 to 300? Whenever you go from 100 to a higher number, the percentage increase is the difference between 100 and the new number: $300 - 100 = 200$. Suppose you quadruple a number ($400 - 100 = 300$); the increase is also the difference between 100 and the new number. Pretty easy, huh?

We'll try one more. When you double a number, the percentage increase is 100%. Use our original formula to calculate this percentage change:

$$\frac{\text{change}}{\text{original number}}$$

If we double a number, in this case, doubling 100, we would get:

$$\frac{100}{100} = 100\%$$

What would the percentage increase be if we went from 7 to 21? It would be 200%. How did we get this? We use the formula—change divided by the original number—or we remember that doubling is a 100% increase and tripling is a 200% increase.

Now I'm going to throw you a curveball. If a number—any number—were to decline by 100%, what number would you be left with? I'd really like you to think about this one.

What did you get? You should have gotten 0. That's right—no matter what number you started with, a 100% decline leaves you with 0.

Try the number 15. Now write down the formula for percentage change:

$$\frac{\text{change}}{\text{original number}}$$

Next, substitute into the formula:

$$-100\% = \frac{?}{15}$$

Since 100% is equal to 1, we have:

$$100\% = \frac{-15}{15}$$

In other words, if the change from the original number 15 is also 15, that means we went from the original number 15 all the way to 0.

This is what happens mathematically. We started with 15, and 15 declined by 100%. What is 100% of 15? It's 15. So we start with 15 and there's a decline of 15. We are left with $15 - 15 = 0$. We can conclude that if any number were to decline by 100%, we'd be left with zero.

SELF-TEST 3

1. A change from 10 to 50 represents a percentage change of how much?
2. A change from 50 to 0 represents a percentage change of how much?
3. A change from 20 to 50 represents a percentage change of how much?
4. A 100% decline from 65 leaves us with how much?

ANSWERS

1. 400% **2.** 100% **3.** 150% **4.** 0

4 Percentage Distribution

Are you eating too much red meat? You may be asking what that has to do with math. First, there are the pluses and the minuses. Red meat is a great source of protein; however, it's also a source of cholesterol and fat. These pluses and minuses will help us illustrate percentage distribution.

Problem 1:

If over the course of a week, you obtained 250 grams of protein from red meat, 150 from fish, 100 from poultry, and 50 from other sources, what percentage of your protein intake came from red meat and what percentage came from each of the other sources?

red meat	250 grams
fish	150 grams
poultry	100 grams
other	50 grams
	550 grams

You now should be able to work this out for yourself. Hint: 550 grams = 100%.

Solution:

$$\text{red meat} = \frac{250}{550} = \frac{25}{55} = \frac{5}{11} = 45.5\% \qquad \begin{array}{r} .4\,5\,4\,5 \\ 11\,\overline{)5.0^{6}0^{5}0^{6}0} \end{array}$$

$$\text{fish} = \frac{150}{550} = \frac{15}{55} = \frac{3}{11} = 27.3\% \qquad \begin{array}{r} .2\,7\,2\,7 \\ 11\,\overline{)3.0^{8}0^{3}0^{8}0} \end{array}$$

$$\text{poultry} = \frac{100}{550} = \frac{10}{55} = \frac{2}{11} = 18.2\% \qquad \begin{array}{r} .1\,8\,1\,8 \\ 11\,\overline{)2.0^{9}0^{2}0^{9}0} \end{array}$$

$$\text{other} = \frac{50}{550} = \frac{5}{55} = \frac{1}{11} = 9.1\% \qquad \begin{array}{r} .09\,09 \\ 11\,\overline{)1.00^{1}00} \end{array}$$

Check:

$$\begin{array}{r} {}^{1} \\ 45.5 \\ 27.3 \\ 18.2 \\ 9.1 \\ \hline 100.1 \end{array}$$

It's always a good idea to run a check of your work. The individual percentage shares should add up to about 100%. In this case, because of rounding, we ended up at slightly more than 100.0%.

Now we'll try another example.

Problem 2:

A family spent $7,000 on food, $4,000 on clothing, $5,500 on shelter, and $3,500 on other miscellaneous items. Find the percentage share of each expenditure.

food	$7,000
clothing	4,000
shelter	5,500
other	3,500
	$20,000

Solution:

$$\text{food} = \frac{7{,}000}{20{,}000} = \frac{7}{20} = .35 = 35\%$$

$$\text{clothing} = \frac{4{,}000}{20{,}000} = \frac{4}{20} = \frac{2}{10} = .2 = 20\%$$

$$\text{shelter} = \frac{5{,}500}{20{,}000} = \frac{55}{200} = \frac{11}{40} = 27.5\%$$

$$40\overline{)11.0} = 4\overline{)1.1^{3}0^{2}0}\ \ (.275)$$

(Note this maneuver we've just executed. We divided 40 by 10 and 11 by 10, making a long division problem into a short one.)

$$\text{other} = \frac{3{,}500}{20{,}000} = \frac{35}{200} = \frac{7}{40}$$

$$40\overline{)7.0} = 4\overline{).7^{3}0^{2}0}\ (.175) = 17.5\%$$

Check:

35.0%
20.0
27.5
17.5
100.0%

SELF-TEST 4

1. A country has 5 million Asians, 8 million Indians, and 2 million blacks. Find the percentage share of each population group.
2. A college had 700 freshmen, 650 sophomores, 600 juniors, and 550 seniors. Find the percentage share of each class.
3. A bank lent $200 million to consumers, $300 million to businesses, and $500 million to the government. Find the percentage share of each type of loan.

ANSWERS

1.

Asians	5 million
Indians	8 million
blacks	2 million
	15 million

$$\text{Asians} = \frac{5}{15} = \frac{1}{3} = 33.3\%$$

$$\text{Indians} = \frac{8}{15} = 15\overline{)8.0^{5}0^{5}0^{5}0}\ (.5333) = 53.3\%$$

$$\text{blacks} = \frac{2}{15} = 15\overline{)2.0^{5}0^{5}0^{5}0}\ (.1333) = 13.3\%$$

Check:
33.3%
53.3
13.3
99.9%

2.

freshmen	700
sophomores	650
juniors	600
seniors	550
	2,500

$$\text{freshmen} = \frac{700}{2{,}500} = \frac{7}{25} = 28\%$$

$$\text{sophomores} = \frac{650}{2{,}500} = \frac{65}{250} = \frac{13}{50} = 26\%$$

$$\text{juniors} = \frac{600}{2{,}500} = \frac{60}{250} = \frac{12}{50} = 24\%$$

$$\text{seniors} = \frac{550}{2{,}500} = \frac{55}{250} = \frac{11}{50} = 22\%$$

Check: 28% + 26 + 24 + 22 = 100%

3.

consumers	$200 million
businesses	300 million
government	500 million
	$1,000 million

$$\text{consumers} = \frac{200}{1{,}000} = 20\%$$

$$\text{businesses} = \frac{300}{1{,}000} = 30\%$$

$$\text{government} = \frac{500}{1{,}000} = 50\%$$

Check: 20% + 30 + 50 = 100%

5 Tipping

One of the greatest social questions of all time is how much to leave as a tip in a restaurant. Etiquette experts Emily Post and Amy Vanderbilt may have hinted at the standard 15%. This percentage might vary, of course, with such factors as quality of service, friendliness of the waitress or waiter, ambiance, quality of the meal, and size of the check (people are less inclined to tip 15% of a $250 check than of a $5 check).

There are plenty of diners who pull out their calculators to determine the exact 15%, but this operation detracts from the atmosphere, especially after a romantic dinner in a candlelit restaurant. So consider your alternatives.

In New York, we often double the tax, which saves us from figuring out just what 15% of the check is. But since the sales tax in New York City is actually 8 1/4%, we tip 16 1/2% without knowing it.

You could try to multiply the check by .15 in your head. But this may be difficult and time-consuming. So let's figure out a fast way of at least approximating 15%.

If your bill comes to $39.82, round it off to $40. How much is 15% of $40? The answer is $6. If you can't calculate .15 × $40 in your head, take 10% of $40. That's $4 (see the box on page 66). How much is 5% of $40? If 10% of $40 is $4, then 5% must be $2. Just add $4 and $2 to arrive at the 15% figure for $40.

If your bill comes to $63.09, how do you calculate a 15% tip? First, round the figure to $63; 10% comes to $6.30, and an additional 5% is $3.15 for a total of $9.45. Just one last word of caution. Try to be real smooth about all this. Don't mutter to yourself, and don't move your lips as you multiply. If you do, you'll lose the whole effect. When Joe DiMaggio patrolled center field for the Yankees, he made every catch look easy.

The next time you leave a tip, gracefully put down exactly the right amount and casually stroll out of the restaurant as if you've been doing 15% in your head all your life.

How to Find 10% of Any Number

How much is 10% of 537? It's 53.7. How much is 10% of 16? It's 1.6. How do you find 10% of any number? Just move the decimal point one place to the left.

Why do we need to find 10% of any number? It can help you calculate tips and prices on sale items. Or if you're going to buy something on credit, you can calculate a 10% down payment. You don't need a calculator to find 10%. You don't even need pencil and paper. All you need to do is move the decimal point one place to the left.

SELF-TEST 5

If you left a 15% tip, how much money would you leave on each of these restaurant checks?

1. $20.00 **2.** $45.00 **3.** $29.50 **4.** $73.95

ANSWERS

1. Ten percent of $20 is $2, and one-half of $2 is $1. So you would leave a $3 tip ($2 + $1).

2. Ten percent of $45 is $4.50. Half of $4.50 is $2.25. So you would leave a tip of $6.75 ($4.50 + $2.25).

3. Be a sport here. Leave 15% of $30, which comes to $4.50. (Ten percent of $30 is $3. Half of $3 is $1.50: $3 + $1.50 = $4.50.)

4. Ten percent of $73.95 is $7.39, rounded to $7.40. Half of $7.40 is $3.70. Your tip would be $11.10 ($7.40 + $3.70).

11 Negative Numbers

In everyday life, we rarely encounter negative numbers, but you never know when they'll crop up. And when they do, watch out! The next time you come across a negative number, you'll be sure to wind up holding the short end of the stick.

1 What Are Negative Numbers?

If you went on a trip and a thief stole your wallet (and you left home without your American Express traveler's checks), you would end up with no money. What could be worse? You get into a friendly poker game, lose all your money, *and* owe your "friends" a couple of thousand dollars. Now that's getting into negative numbers.

Or take the case of Primo Carniera, a boxer, who decided to become a professional wrestler back in the 1950s. Why did Mr. Carniera change jobs? Well, it seems that former managers and trainers owned 115 percent of him. If Carniera got a $100,000 purse for a fight, after he paid out his managers' and trainers' shares, he ended up with minus $15,000. So every time the man fought, he lost money. Oh well, boxing's loss was wrestling's gain.

The most popular negative number is business losses. If your firm has sales of $500,000 and costs of $550,000, it has lost $50,000. If we think of a loss as a negative profit, then your business made a profit of − $50,000.

It may help to count down from positive to negative numbers. Positive numbers are ordinary everyday numbers: 5, 2, 16, 37, 9, 104. We could put + signs in front of them: +5, +2, +16, +37, +9, +104, but usually we don't bother. The only time we use these signs is when there are some negative numbers close by.

Let's start a countdown. I'll provide the first three numbers and you supply the next three.

Problem 1:

4, 3, 2, ______, ______, ______

Solution:

What did you get? You got 1 for the next number? So far, so good. What's the next number? It's 0. And the one after 0? It's −1.

Here's another set.

Problem 2:

2, 1, 0, −1, ______, ______, ______

Solution:

I'll bet you wrote −2, −3, −4. Good, you're really getting into negative numbers now.

2 Adding Negative Numbers

Negative numbers can be added, subtracted, multiplied, and divided just as positive numbers can. First we'll do addition.

Add −3 and −5. Did you get −8? That's right. If you lost $3 to your poker-playing friends and then lost another $5, you'd be $8 in the hole.

Add +4 and −6. Did you get −2? I hope so. You had been ahead by 4 and then lost 6, so now you're behind by 2.

Let's try adding −3, +4, −1, −5, and +2. And the envelope, please. The answer is −3. If we add the pluses we get +6; if we add the minuses we get −9. Adding +6 and −9, we get −3.

3 Subtracting Negative Numbers

Ready for some subtraction? Then here goes. Subtract −6 from −2. Your answer should be +4. That's right! Imagine that you did your income tax return and figured out that you owed the government $2,000, but then the Internal Revenue Service informed you that you had made a \$6,000 mistake on your tax return and that they were sending you a \$4,000 refund.

Here's another problem. Subtract -3 from -10. What did you get? The right answer is -7. Here's the orthodox way of doing this: $-10-(-3)=-10+3=-7$. In other words, subtracting a -3 is really the same thing as adding a $+3$.

And finally, one last problem. Subtract -9 from -5. You get $-5-(-9)=-5+9=+4$.

SELF-TEST 1

1. Counting down: 5, 4, 3, ______, ______, ______, ______, ______
2. Counting up: -5, -4, -3, ______, ______, ______, ______, ______
3. Add -4 and -3.
4. Add -3, -1, $+2$, -6, and -2.
5. Add -5, $+6$, -2, -4, and $+7$.
6. Subtract -2 from $+9$.
7. Subtract -6 from -4.
8. Subtract $+5$ from -2.

ANSWERS

1. 2,1,0, -1, -2 **2.** -2, -1, 0, $+1$, $+2$ **3.** -7 **4.** -10 **5.** $+2$ **6.** 11 **7.** $+2$ **8.** -7

If subtracting negative numbers is still not your strong suit, then take a look at the box on p. 70 and then go on to frame 4.

4 Multiplying Negative Numbers

Ready to move up to a classier operation? Surprisingly, multiplying and dividing negative numbers is relatively easy. What's that? You're not convinced?

OK, there is a catch. When you multiply two numbers, if just one of them has a negative sign, the answer is negative. Multiply -3×4. You should have gotten -12. Multiply 6×-3. Did you get -18?

When you multiply two positive numbers, you get a positive product. You've been doing that all your life. And when you multiply two negative numbers, you also get a positive product: $-3 \times -2 = +6$; $-5 \times -3 = +15$, or just plain 15.

Do you think you've gotten the hang of it? There's only one way to find out.

Find the product of -7×2. It's -14. A negative times a positive gives us a negative.

Find the product of -3×-8. It's 24. Two negatives give us a positive.

And finally, how much is 5×6? It's 30. Two positives give us a positive.

Subtracting a Negative Is Always a Plus

It may help to once again think of a negative number as a debt. Subtracting that debt is the same as adding that debt to your wealth. For instance, if you're $7 in debt and your creditor tells you to forget about the debt, you can state what happened this way:

$-\$7 - (-\$7) = -\$7 + \$7 = 0$

So now you're out of debt. When we subtract a minus, it's the same as adding a plus.

Problem:

How much is -2 subtracted from -6?

Solution:

$-6 - (-2) = -6 + 2 = -4$

Problem:

How much is 5 minus a -4?

Solution:

$5 - (-4) = 5 + 4 = 9$

Find the products:

1. -1×-2 **2.** $-8 \times +7$ **3.** 6×9 **4.** 4×-7

5. 9×-9 **6.** -5×-10 **7.** 8×-8

ANSWERS

1. +2 **2.** −56 **3.** +54 **4.** −28 **5.** −81 **6.** +50 **7.** −64

Note: If you are still having any trouble doing multiplication, you should definitely review the multiplication table (see Table 2.2). And it would probably be a good idea to take each of the self-tests in chapters 3 through 5. Remember the old song by Diana Ross and the Supremes, “You Can’t Hurry Love”? Well, the song we sing here is, you can’t hurry math.

5 Division with Negative Numbers

At last we come to division with negative numbers—the moment we’ve all been waiting for. But first, we’ll review our rules.

Division involving two positive numbers yields a positive quotient. Two negatives give us a positive quotient. And finally, when one number is positive and one is negative, the quotient will be negative.

Is all this clear? Just remember, if both numbers have the same sign, the quotient is *positive*. If they have different signs, it’s *negative*.

Divide -3 into 6. The answer is -2.
Divide -5 into -25. The answer is $+5$.
Divide 16 by -4. The answer is -4.
Divide -42 by 6. The answer is -7.
How much is $56 \div -8$? The answer is -7.
How much is $-63 \div -7$? The answer is $+9$.

SELF-TEST 3

1. Divide -8 into 64.
2. Divide 7 into -35.
3. Divide -81 by -9.
4. Divide -80 by 10.
5. How much is $-36 \div 4$?
6. How much is $72 \div -9$?

1. -8 2. -5 3. $+9$ 4. -8 5. -9 6. -8

12 The Isolation of x

Throughout this chapter, we'll be using equations. What's an equation? It's a mathematical statement that tells us that what is on one side of an equal sign (=) is equal to what's on the other side. For example, $3+4=7$.

1 The Care and Treatment of Equations

Now suppose we decide to add 3 to the left side of the previous equation: $3+4+3$. What must we do? That's right! We must add 3 to the right side: $7+3$.

Let's try another equation: $5+4=9$. If we subtracted 2 from the left side: $5+4-2$, what comes next? You guessed it! We must subtract 2 from the right side: $9-2$.

It follows that if the right side were multiplied by a certain number, we'd have to multiply the left side by that same number. Finally, if one side is divided by 8, for example, the other side must also be divided by 8.

So what we do to one side of the equation, we must do to the other side. Why? Because the two sides, by definition, are always equal.

2 Isolating *x*

In algebra, x is often used to designate the unknown quantity—hence the quantity we are seeking to find. It follows that we would want to isolate x and find out exactly what it is. By isolating x, I mean we put it all alone on one side of the equal sign.

3 Addition and Subtraction with x

Problem 1:

If $x - 6 = 2$, how much is x?

Solution:

$$x - 6 = 2$$
$$x - 6 + 6 = 2 + 6$$
$$x = 8$$

To isolate x, we added 6 to both sides of the equation.

Now let's check our work. Go back to the original equation, $x - 6 = 2$ and substitute your answer, 8, for x. What do you get?

$$x - 6 = 2$$
$$8 - 6 = 2$$
$$2 = 2$$

So it checks.

Problem 2:

If $x - 4 = 9$, how much is x?

Solution:

$$x - 4 = 9$$
$$x = 13$$

Here we added 4 to both sides.

Check your work here, by substituting your answer into the original equation. Does it check out? I'll do it once more:

$$x - 4 = 9$$
$$13 - 4 = 9$$
$$9 = 9$$

Problem 3:

If $x + 3 = 4$, how much is x?

Solution:

$$x+3=4$$
$$x+3-3=4-3$$
$$x=1$$

Here we subtracted 3 from both sides to isolate x.

Problem 4:

If $x+6=9$, how much is x?

Solution:

$$x+6=9$$
$$x+6-6=9-6$$
$$x=3$$

Still a little confused about whether to add or subtract? Then remember, we want to isolate x. In the first problem, $x-6=2$, we had to get rid of the -6. We did that by adding $+6$ to both sides. In each of the subsequent problems, we isolated x by getting rid of the number that was on x's side of the equation. And what we did on one side of the equation, we did on the other side as well.

SELF-TEST 1

Find x in each of these problems:

1. $x-8=17$ **2.** $x-4=5$ **3.** $x+4=12$ **4.** $x+10=15$

ANSWERS

1. $x=25$ **2.** $x=9$ **3.** $x=8$ **4.** $x=5$

4 Multiplication and Division with *x*

Problem 1:

We'll begin with an equation: $2x=50$. How much is x?

Solution:

$$2x=50$$
$$x=25$$

How do we know that x is 25? It's elementary, my dear Watson. We divide both sides of the equation by 2 to isolate x.

Run your check on the answer by substituting it into the original equation.

Check:

$$2x = 50$$
$$2 \times 25 = 50$$
$$50 = 50$$

Here's another one.

Problem 2:

If $3x = 12$, how much is x?

Solution:

$x = 4$

Problem 3:

Find x when $x/3 = 2$.

Solution:

$$\frac{x}{3} = 2$$
$$\frac{(\cancel{3})x}{\cancel{3}} = (3) \times 2$$
$$x = 6$$

We multiplied both sides of the equation by 3.

Problem 4:

Solve for x when $x/4 = 10$.

Solution:

$$\frac{x}{4} = 10$$
$$\frac{(\cancel{4})x}{\cancel{4}} = (4)x10$$
$$x = 40$$

SELF-TEST 2

Find x in each of these problems:

1. $4x = 8$
2. $6x = 15$
3. $\frac{x}{2} = 9$
4. $\frac{x}{5} = 4$

ANSWERS 1. $x=2$ 2. $x=2\frac{1}{2}$ 3. $x=18$ 4. $x=20$

5 Decimals with *x*

Now that you've been around the track a few times, you may be ready to have a few decimals tossed at you. You're still after just one thing—you want to get *x* all alone.

Problem 1:

If $.2x=8$, how much is x?

Solution:

$$.2x=8$$

$$\frac{.2x}{.2}=\frac{8}{.2}$$

$$x=40$$

To isolate *x*, we divided both sides by .2. What we do to one side of the equation, we must do to the other side.

Does this answer check out? There's only one way to find out.

Check:

$$.2x=8$$

$$.2\times 40=8$$

$$8=8$$

Problem 2:

If $2.5x=17$, how much is x?

Solution:

$$2.5x=17$$

$$\frac{2.5x}{2.5}=\frac{17}{2.5} \qquad 2.5\overline{)17.0} \qquad \begin{array}{r} 6.8 \\ 25\overline{)170.0} \end{array}$$

$$x=6.8$$

Check:

$$\begin{array}{r} 6.8 \\ \times 2.5 \\ \hline 3\,40 \\ 13\,6 \\ \hline 17.00 \end{array}$$

We'll work out one more.

Problem 3:

Find x when $4.3x = 10$.

Solution:

$$4.3x = 10$$

$$\frac{4.3x}{4.3} = \frac{10}{4.3} \qquad 4.3\,\overline{)10} \qquad \begin{array}{r} 2.3 \\ 43\,\overline{)100.0} \\ -86 \\ \hline 140 \\ -129 \\ \hline 11 \end{array}$$

$$x = 2.3$$

SELF-TEST 3

Find x in each of these problems:

1. $.4x = 9$
2. $.9x = 12$
3. $1.8x = 2$
4. $3.5x = 15$

ANSWERS

1. $\frac{.4x}{.4} = \frac{9}{.4}$ $x = 22.5$
2. $\frac{.9x}{.9} = \frac{12}{.9}$ $x = 13.3$
3. $\frac{1.8x}{1.8} = \frac{2}{1.8}$ $x = 1.1$
4. $\frac{3.5x}{3.5} = \frac{15}{3.5}$ $x = 4.3$

6 Negative Numbers

Can x ever be a negative number? It sure can. Suppose that x represents your losses at the track this year. Or the money you spent on lottery tickets.

Problem 1:

If $x + 7 = 3$, how much is x?

Solution:

$$x + 7 = 3$$
$$x + 7 - 7 = 3 - 7$$
$$x = -4$$

What we do here is follow the same procedure we used in the first section of this chapter to isolate x. If you're at all uncertain about this procedure, please return to frame 3.

Problem 2:

If $x + 10 = 2$, how much is x?

Solution:

$$x + 10 = 2$$
$$x + 10 - 10 = 2 - 10$$
$$x = -8$$

Problem 3:

Find x if $x + 15 = -4$.

Solution:

$$x + 15 = -4$$
$$x + 15 - 15 = -4 - 15$$
$$x = -19$$

I hope you're still checking your answers.

Check:

$$x + 15 = -4$$
$$-19 + 15 = -4$$
$$-4 = -4$$

Problem 4:

Find x if $x - 2 = -5$.

Solution:

$$x - 2 = -5$$
$$x - 2 + 2 = -5 + 2$$
$$x = -3$$

Problem 5:

Find x when $x - 6 = -7$.

Solution:

$$x - 6 = -7$$
$$x - 6 + 6 = -7 + 6$$
$$x = -1$$

If negative numbers make you nervous, then you need to review chapter 11. After you've done that, return to frame 6.

If you did not have any difficulty with these problems, then you shouldn't have any difficulty with those in Self-Test 4.

SELF-TEST 4

Find x in each of these problems:

1. $x+4=-8$
2. $x+9=2$
3. $x-7=-10$
4. $x+5=4$
5. $x+3=2$
6. $x-10=-14$

ANSWERS

1. $x=-12$ 2. $x=-7$ 3. $x=-3$ 4. $x=-1$ 5. $x=-1$ 6. $x=-4$

7 Combination Problems with x

The problems in this section combine the various things we've covered in this chapter—addition, subtraction, multiplication, division, decimals, and negative numbers.

Problem 1:

If $2x+5=10$, solve for x.

Solution:

$$2x+5=10$$
$$2x=5$$
$$x=\frac{5}{2}$$
$$x=2\frac{1}{2}$$

What we've done here is combine two operations to isolate x. First we got rid of the 5 by subtracting it from both sides of the equation. This left us with $2x$. To get x, we divided both sides by 2.

Problem 2:

If $5x-3=9$, solve for x.

Solution:

$$5x-3=9$$
$$5x=12$$
$$x=\frac{12}{5}$$

$$x = 2\frac{2}{5}$$

Problem 3:

Find x when $x/3 + 2 = 5$.

Solution:

$$\frac{x}{3} + 2 = 5$$
$$\frac{x}{3} = 3$$
$$x = 9$$

Problem 4:

How much is x when $2/3x - 1 = 8$?

Solution:

$$\frac{2}{3}x - 1 = 8$$
$$\frac{2}{3}x = 9$$
$$\frac{3}{2} \times \frac{2}{3}x = \frac{3}{2} \times \frac{9}{1}$$
$$x = \frac{27}{2}$$
$$x = 13.5$$

Problem 5:

Find x when $6x + 11 = 3$.

Solution:

$$6x + 11 = 3$$
$$6x = -8$$
$$x = -1\frac{1}{3}$$

Problem 6:

When $3/5x - 2 = -10$, how much is x?

Solution:

$$\frac{3}{5}x-2=-10$$

$$\frac{3}{5}x=-8$$

$$\frac{5}{3}\times\frac{3}{5}x=\frac{5}{3}\times\frac{-8}{1}$$

$$x=\frac{-40}{3}$$

$$x=-13\frac{1}{3}$$

Problem 7:

If $.5x-1=-6$, find x.

Solution:

$$.5x-1=-6$$

$$.5x=-5$$

$$x=-10$$

Problem 8:

When $2.8x+12=3$, how much is x?

Solution:

$$2.8x+12=3$$

$$2.8x=-9$$

$$\frac{2.8x}{2.8}=-\frac{9}{2.8}$$

$$x=-2.8\overline{)9}$$

$$-28\overline{)90}$$

$$\begin{array}{r} -3.2 \\ -14\overline{)45.30} \end{array}$$

$$x=-3.2$$

SELF-TEST 5

Find x in each of these problems:

1. $4x+7=15$ **2.** $2x-7=13$ **3.** $\frac{x}{5}+4=19$ **4.** $\frac{3}{4}x-1=-6$

5. $9x+1=0$ **6.** $.7x+3=15$ **7.** $.4x+12=3$ **8.** $3.1x-9=-1$

ANSWERS

1. $4x + 7 = 15$
$4x = 8$
$x = 2$

2. $2x - 7 = 13$
$2x = 20$
$x = 10$

3. $\frac{x}{5} + 4 = 19$
$\frac{x}{5} = 15$
$x = 75$

4. $\frac{3}{4}x - 1 = -6$
$\frac{3}{4}x = -5$
$3x = -20$
$x = -6\frac{2}{3}$ or -6.7

5. $9x + 1 = 0$
$9x = -1$
$x = -\frac{1}{9}$

6. $.7x + 3 = 15$
$.7x = 12$
$.7\overline{)12}$
$7\overline{)12^{5}0.^{1}0^{3}0}$ = 1 7. 1 4
$x = 17.1$

7. $.4x + 12 = 3$
$.4x = -9$
$-.4\overline{)9}$
$-4\overline{)9^{1}0.^{2}0}$ = $-2\,2.\,5$
$x = -22.5$

8. $3.1x - 9 = -1$
$3.1x = 8$
$3.1\overline{)8}$

$$
\begin{array}{r}
2.58 \\
31\overline{)80.00} \\
\underline{-62^{xx}} \\
18\,0 \\
\underline{-15\,5} \\
2\,50 \\
\underline{-2\,48}
\end{array}
$$

$x = 2.6$

13 Exponents and Square Roots

How much is $2 \times 2 \times 2 \times 2$? The answer is 16. Another way of expressing $2 \times 2 \times 2 \times 2$ is 2^4, or 2 to the fourth power. In algebra, which we started in the last chapter, this concept is extremely important.

1 Exponents: The Powers of x

What is an exponent? The terms x^2 and x^3 have exponents. They're the little numbers to the right and just above each x. The first term, x^2, is read x squared; the second, x^3, is read x cubed. With that introduction, I won't call them exponents anymore—just squares and cubes.

Oh, I forgot to explain what a square is and what a cube is. A square is a number times itself: $3^2 = 3 \times 3$; $5^2 = 5 \times 5$. And x^2? That's x times x. A cube is a number times itself twice. So $3^3 = 3 \times 3 \times 3$; $5^3 = 5 \times 5 \times 5$. And $x^3 = x$ times x times x.

Problem 1:

How much is x^2 if $x = 4$?

Solution:

$x^2 = 4^2 = 4 \times 4 = 16$

Problem 2:

How much is x^2 if $x = 6$?

Solution:

$x^2 = 6^2 = 6 \times 6 = 36$

Next we'll do cubes.

Problem 3:

How much is x^3 if $x = 2$?

Solution:

$x^3 = 2^3 = 2 \times 2 \times 2 = 8$

Problem 4:

How much is x^3 if $x = 3$?

Solution:

$x^3 = 3^3 = 3 \times 3 \times 3 = 27$

Now, while we're hot, we might as well talk about x^1, or x to the first power. That's the same as x. For example, if x is 7, how much is x^1? The answer is 7. And if $x^1 = 4$, how much is x? It is 4.

Finally we have x^0. Anything to the zero power equals one. So, x to the 0 power is equal to 1. That comes up occasionally in math, but you won't have to deal with it in this book (except in Self-Test 1). Also, I won't bother you with x^{-1}, x^{-2}, x^{-3}, and so forth.

SELF-TEST 1

1. If $x = 3$, how much is x^4?
2. If $x = 5$, how much is x^2?
3. If $x = 2$, how much is x^3?
4. If $x = 8$, how much is x^1?
5. If $x = 10$, how much is x^0?
6. If $x = 4$, how much is
   a. x^0 b. x^1 c. x^2 d. x^3 e. x^4

ANSWERS

1. $3 \times 3 \times 3 \times 3 = 81$ **2.** $5 \times 5 = 25$ **3.** $2 \times 2 \times 2 = 8$ **4.** 8 **5.** 1

6. a. 1 b. 4 c. $4 \times 4 = 16$ d. $4 \times 4 \times 4 = 64$ e. $4 \times 4 \times 4 \times 4 = 256$

The Powers That Be

Here's a summary of the powers of 1:

$1^0 = 1$
$1^1 = 1$
$1^2 = 1$
$1^3 = 1$
$1^4 = 1$

And here's a summary of the powers of 10:

$10^0 = 1$
$10^1 = 10$
$10^2 = 100$
$10^3 = 1{,}000$
$10^4 = 10{,}000$

2 Square Roots

We're going to come dangerously close to using a calculator, but we'll manage somehow without one. I'll tell you right up front, almost every calculator can do square roots.

What's a square root? The square root of x^2 is x. What's the square root of 4? What number times itself gives us 4? The answer is ± 2. So ± 2 is the square root of 4. We denote square root by the sign $\sqrt{\ }$. For example, $\sqrt{4}$ means the square root of 4.

Find $\sqrt{9}$. What number times itself is 9? The answer is ± 3. The square root of 9 is ± 3. Why do we say ± 3? Why not just 3 or $+3$? With the help of frame 4 in chapter 11 and a little logic, we can answer that question.

Frame 4 of chapter 11 dealt with the multiplication of negative numbers. When two negative numbers are multiplied, their product is positive. For example, $-3 \times -3 = +9$, or 9. To find $\sqrt{9}$, we need to look for the number that, multiplied by itself, yields a product of 9. That number could be either $+3$ or -3.

Now we'll do another one. How much is the square root of 16? That's right. It's ± 4.

Figure this one out—if $x^2 = 36$, how much is x? The answer is ± 6.

You may have noticed that we didn't try to figure out the square roots of just any old numbers. The numbers we chose—4, 9, 16, and 36—all had nice whole number square roots. What about other numbers, such as 5, 6, 7, or 8? These numbers all have squares roots of more than ± 2, but less than ± 3. How do we know that? Because the square root of 4 is ± 2 and the square root of 9 is ± 3.

And how would we find the square roots of these numbers? The chances are, you'll never have to. Some texts (not this one) print tables of square roots. Some math texts (not this one) show you how to figure out square roots. Of course, if you really must find the square root of some number, you have our permission to use the c-word, which, we're forced to disclose, is calculator.

There is a way of approximating a square root through multiplication. Let's take the square root of 6. We already know that the square root of 4 is ± 2 and the square root of 9 is ± 3. So the square root of 6 is somewhere between ± 2 and ± 3. It's probably more than ± 2.1 or ± 2.2 and less than ± 2.8 or ± 2.9.

Let's try ± 2.3. Does $\pm 2.3 \times \pm 2.3$ gives us 6? Let's find out.

$$\begin{array}{r} \pm 2.3 \\ \times \pm 2.3 \\ \hline 69 \\ 46 \\ \hline 5.29 \end{array}$$

So, we'll try ± 2.4.

$$\begin{array}{r} \pm 2.4 \\ \times \pm 2.4 \\ \hline 96 \\ 48 \\ \hline 5.76 \end{array}$$

We know we're getting close. We can say that the square root of 6 is slightly more than ± 2.4. If we really want to get closer, we can try $\pm 2.45 \times \pm 2.45$.

SELF-TEST 2

Find the square root of

1. 100 **2.** 64 **3.** 49 **4.** 81

How much is *x* if

5. $x^2 = 9$ **6.** $x^2 = 25$ **7.** $x^2 = 16$ **8.** $x^2 = 4$

ANSWERS

1. ± 10 **2.** ± 8 **3.** ± 7 **4.** ± 9

5. ± 3 **6.** ± 5 **7.** ± 4 **8.** ± 2

3 Advanced Problems

Problem 1:

If $x^2 + 4 = 20$, find x.

Solution:

$x^2 + 4 = 20$
$x^2 = 16$
$x = \pm 4$

Problem 2:

If $2x^2 = 50$, find x.

Solution:

$2x^2 = 50$
$x^2 = 25$
$x = \pm 5$

Problem 3:

If $3x^2 - 5 = 7$, find x.

Solution:

$3x^2 - 5 = 7$
$3x^2 = 12$
$x^2 = 4$
$x = \pm 2$

Problem 4:

If $x = 10$, how much is $x^2 - 5$?

Solution:

$x^2 - 5 =$
$10^2 - 5 =$
$100 - 5 = 95$

Problem 5:

If $x = 3$, how much is $2x^2 + 1$?

Solution:

$$2x^2 + 1 =$$
$$2 \times 3^2 + 1 =$$
$$2 \times 9 + 1 =$$
$$18 + 1 = 19$$

Problem 6:

If $x = 8$, how much is $3x^2 + 2x - 3$?

Solution:

$$3x^2 + 2x - 3 =$$
$$(3 \times 8 \times 8) + (2 \times 8) - 3 =$$
$$192 + 16 - 3 = 205$$

SELF-TEST 3

How much is x?

1. $2x^2 - 90 = 72$ **2.** $10 + 3x^2 = 37$ **3.** $5x^2 + 1 = 126$ **4.** $7 - 4x^2 = -57$

Solve each of these problems:

5. If $x = 8$, how much is $150 - 2x^2 + x$?

6. If $x = 1$, how much is $2x^2 - 1$?

7. If $x = 7$, how much is $14 - x^2$?

8. If $x = 10$, how much is $3x^2 + 4$?

ANSWERS

1. $2x^2 = 162$
$x^2 = 81$
$x = \pm 9$

2. $3x^2 = 27$
$X^2 = 9$
$x = \pm 3$

3. $5x^2 = 125$
$x^2 = 25$
$x = \pm 5$

4. $-4x^2 = -64$
$4x^2 = 64$
$x^2 = 16$
$x = \pm 4$

5. $150 - (2 \times 8 \times 8) + 8 = 150 - 128 + 8 = 30$

6. $(2 \times 1 \times 1) - 1 = 2 - 1 = 1$

7. $14 - (7 \times 7) = 14 - 49 = -35$

8. $(3 \times 10 \times 10) + 4 = 300 + 4 = 304$

14 Ratios and Proportions

When someone says you're blowing things way out of proportion, what does that mean? Way out of proportion to what? Are you the kind of person who makes mountains out of molehills? As we study this chapter, we will come to understand the meaning and implications of ratios and proportions.

1 Ratios

Before we can discuss proportions, we must examine ratios. If someone says, I'll give you three to one on the Mets over the Yankees in the World Series, what does that person mean?

Three to one, or 3:1, is a ratio. If someone gives you 3 to 1, it's his $3 to your $1.

We use ratios all the time. A map may have a scale of 1 inch to a mile. If two points on the map are 3 inches apart, they're actually 3 miles apart.

How about a Bloody Mary that's mixed 1 to 1? Too strong? How about 3 to 1 (3 parts tomato juice to 1 part vodka)?

There are 12 inches in a foot, or 12:1. There are two pints in a quart, or 2:1. And there are four quarts in a gallon, or 4:1. The two vertical dots, :, clearly stand for the word "to." So we can express a ratio as 3:1, or 3 to 1.

Problem 1:

What is the Chicago Bears' ratio of wins to losses if they won 14 of 16 games?

Solution:

14:2 = 7:1

Why not leave it at 14:2? There's nothing wrong with 14:2 mathematically, but by convention, ratios are reduced to their simplest form. The advantage of this form is that it aids us in comparing different ratios. It's a lot easier to compare 3:1 than 192:64.

Problem 2:

What is the ratio of ounces to pounds? (Hint: It helps to know that there are 16 ounces in every pound!)

Solution:

16:1

SELF-TEST 1

1. If someone offers you 8 to 1 odds on the heavyweight championship fight, how much money would you have to put up to win $40?
2. What is the ratio of inches per yard?
3. What is the ratio of quarts to ounces?
4. If a map has a scale of 1 1/2 inches to the mile and if a distance between two points on the map is 9 inches apart, how much is that distance in actual miles?
5. If an employee was out sick on 6 of 96 work days, what is his ratio of sick days to days worked?
6. If a runner worked out 32 of the last 40 days, what is her ratio of days she worked out to days she did not work out?

1. $5 **2.** 36:1 **3.** 1:32 **4.** 6 miles **5.** 6:90 = 1:15 **6.** 32:8 = 4:1

2 Proportions

We're ready for proportions. What exactly is a proportion? It's a statement that two ratios are equal. No mountains made out of molehills here!

Here's a proportion: 1 is to 2 as 3 is to 6, so $1:2 = 3:6$. In this case, 1 has the same relation to 2 that 3 has to 6. And what exactly is that relation? What do you think? You're right: 1 is half the size of 2, and 3 is half the size of 6. Or, alternately, 2 is twice the size of 1, and 6 is twice the size of 3.

Now for a little algebra.

Problem 1:

$1:4 = 3:x$. Find x.

Solution:

$x = 12$. After all, $4 = 4 \times 1$, and $12 = 4 \times 3$. Right answer, sound logic, but we didn't use algebra. What's the point of doing all that algebra in the last two chapters if we never use it?

In order to use it, we need a formula. And here it comes: The product of the means (the inside numbers) equals the product of the extremes (the outside numbers). In simple English, multiply the inside numbers, multiply the outside numbers, and slap down an equal sign between them.

So, we have $1:4 = 3:x$. Multiplying the means, $4 \cdot 3$, gives us 12. Multiplying the extremes, $1 \cdot x$, gives us $1x$, or x. So we end up with $12 = x$.

Let's do another one while we're hot.

Problem 2:

Find x when $2:x = 5:15$.

Solution:

$$2:x = 5:15$$
$$2(15) = 5x$$
$$30 = 5x$$
$$6 = x$$

How about this one?

Problem 3:

3 is to 5 as 9 is to what?

Solution:

The "what" we're looking for, we'll call x.

$$3{:}5 = 9{:}x$$
$$3(x) = 5(9)$$
$$3x = 45$$
$$x = 15$$

Now try this.

Problem 4:

4 is to what as 9 is to 30?

Solution:

$$4{:}x = 9{:}30$$
$$4(30) = 9(x)$$
$$120 = 9x$$
$$40 = 3x$$
$$13\frac{1}{3} = x$$

Problem 5:

If the ratio of your rent to your monthly income is 1:4, how much is your rent if your income is $600?

Solution:

Let rent $= x$

$$1{:}4 = x{:}\$600$$
$$4x = \$600$$
$$x = \$150$$

Problem 6:

If the ratio of Adam's age to his grandmother's age is 2:11 and Adam is 12, how old is his grandmother?

Solution:

Let grandmother's age $= x$

$$2{:}11 = 12{:}x$$
$$2x = 132$$
$$x = 66$$

Are you ready for some apples and oranges?

Problem 7:

If 4 apples cost 39 cents, how much would 13 apples cost?

Solution:

Let $x =$ the cost of 13 apples

$4{:}39 \text{ cents} = 13{:}x$

$4x = 507 \text{ cents}$

$x = 126.75 \text{ cents}$

$x = \$1.27$

Problem 8:

If 7 oranges cost $1.05, how many oranges could be purchased for $3.00?

Solution:

Let $x =$ the number of oranges that could be purchased for $3

$7{:}\$1.05 = x{:}\$3.00$

$\$21 = \$1.05x$

$20 = x$

Problem 9:

A plane travels 2,000 miles in 4 1/2 hours. How long would it take that plane to travel 3,000 miles at that same rate of speed?

Solution:

Let $x =$ length of time it would take the plane to travel 3,000 miles

$4.5{:}2{,}000 = x{:}3{,}000$

$13{,}500 = 2{,}000x$

$13.5 = 2x$

$6.75 = x$

Problem 10:

If the ratio of Asians to non-Asians in San Francisco is 1:6, how many Asians live in San Francisco if the city's population is 770,000?

Solution:

Let $x =$ number of Asians living in San Francisco

$1{:}7^* = x{:}770{,}000$

$770{,}000 = 7x$

$110{,}000 = x$

*We get 1:7 by adding Asians to non-Asians. $1{:}(6 + 1) = 1{:}7$.

SELF-TEST 2

1. If the Boston Celtics win 63 of 81 games, find their win-loss ratio.
2. Solve for x: 9 is to 12 as what is to 40?
3. Solve for x: 6 is to 5 as 48 is to what?
4. If you spend $1 out of every $10 of income on medical bills and if you paid $1,171 in medical bills, how much was your income?
5. $15:2 = x:10$. Find x.
6. $x:3 = 4:9$. Find x.
7. If 8 bananas cost $1.50, how much would 72 bananas cost?
8. If 2 pineapples cost $3.79, how many pineapples could be purchased for $22.74?
9. If you can drive 200 miles in 3 3/4 hours, how long would it take you to drive 500 miles at the same rate?
10. If 1,400,000 of Chicago's 3,100,000 people are black, what is Chicago's ratio of blacks to non-blacks?
11. If the ratio of Hispanics to non-Hispanics in Houston is 1:12, how many Hispanics would there be if Houston had a population of 2 million?

1. $63:18 = 7:2$ **2.** 30 **3.** 40 **4.** $11,710 **5.** 75 **6.** $x = 1\ 1/3$
7. $13.50 **8.** 12 **9.** 9.375 hours **10.** 14:17 **11.** 153,846

15 Finding the Areas of Rectangles and Triangles

How big is a lot that is 2,000 square feet? If carpeting cost $13 a yard, how much would it cost you to carpet a room that was 18 feet by 24 feet? Wen you have completed this chapter, you'll be able to answer all kinds of questions involving the areas of rectangles and triangles.

1 Areas of Rectangles

A rectangle is a two-dimensional box. Three rectangles are pictured. Rectangle A is a square. Notice that its sides are of equal length. Rectangle B's length is a little longer than its width. Rectangle C's length is much longer than its width.

The area of a rectangle is its length times its width.

Problem 1:

How much is the area of Rectangle A?

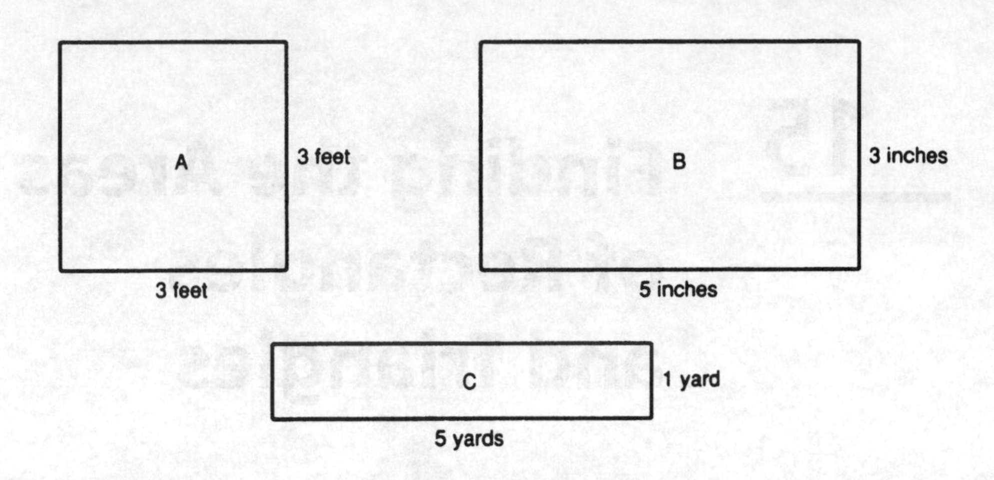

Solution:

Area = length × width
= 3 feet × 3 feet
= 9 square feet

It is not enough to say that the area of Rectangle A is 9 or even 9 feet. Technically it is 9 square feet.

Problem 2:

How much is the area of Rectangle B?

Solution:

Area = length × width
= 5 inches × 3 inches
= 15 square inches

Problem 3:

And how much is the area of Rectangle C?

Solution:

Area = length × width
= 5 yards × 1 yard
= 5 square yards

And now for something a little bit different.

Problem 4:

How many square inches are there in a square foot? Draw a rectangle (which is square). How much is its length and width? Hint: Plot everything in inches.

Solution:

We see from the following figure that the length and width of the square are both 12 inches. Therefore,

Area = length × width
= 12 inches × 12 inches
= 144 square inches

12 inches

12 inches

Algebraically we *could* say that the formula for the area of a square is its side squared, or s^2. I'd rather treat the square as just another rectangle so that we can stick to one formula for area: length × width.

Here's a similar question.

Problem 5:

How many square feet are there in a square yard?

Solution:

We see from the following figure that the length and width of the square are both 3 feet. Therefore,

Area = length × width
= 3 feet × 3 feet
= 9 square feet

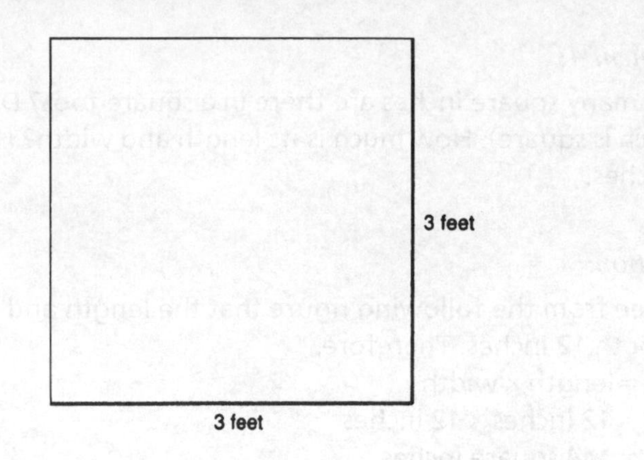

SELF-TEST 1

1. What is the area of a square that has sides of 6 inches?
2. What is the area of a rectangle that is 5 inches long and 4 inches wide?
3. What is the area of a rectangle that is 7 feet long and 3 feet wide?

ANSWERS

1. 6 inches × 6 inches = 36 square inches
2. 5 inches × 4 inches = 20 square inches
3. 7 feet × 3 feet = 21 square feet

2 When We Use Area

Now, we'll do a few problems that you might be likely to encounter every day.

Problem 1:

A commercial building is renting for $10 per square foot. How much would it cost to rent an office that is 50 feet × 100 feet?

Solution:

Area = length × width
= 100 feet × 50 feet
= 5,000 square feet

5,000 square feet × $10 = $50,000

Problem 2:

You buy a carpet that sells for $12 per square yard. If your room is 12 feet by 9 feet, how much will you spend?

Solution:

$$\begin{aligned} \text{Area} &= \text{length} \times \text{width} \\ &= 9 \text{ feet} \times 12 \text{ feet} \\ &= 108 \text{ square feet} \end{aligned}$$

There are 9 square feet in a square yard (3 feet × 3 feet = 9 square feet)—108/9 = 12 square yards, and 12 × \$12 = \$144.

Problem 3:

Manhattan's Central Park is 2 1/2 miles long and 1/2 mile wide. What is its area?

Solution:

$$\begin{aligned} \text{Area} &= \text{length} \times \text{width} \\ &= \frac{5}{2} \times \frac{1}{2} \\ &= \frac{5}{4} = 1\frac{1}{4} \text{ square miles} \end{aligned}$$

Problem 4:

What is the length of a rectangular lot that is 2,000 square feet and is 40 feet wide?

Solution:

$$\begin{aligned} \text{Area} &= \text{length} \times \text{width} \\ 2{,}000 \text{ square feet} &= \text{length} \times 40 \text{ feet} \\ 50 \text{ feet} &= \text{length} \end{aligned}$$

SELF-TEST 2

1. What is the area in square feet of a room that is 3 yards long and 2 yards wide?
2. If land were selling for \$100 per square foot, how much would you have to pay for a square-shaped piece of land 80 feet long?
3. If a room is 30 feet by 30 feet, how much would it cost for wall-to-wall carpeting at \$15 per yard?
4. If a room were 3,000 square feet with a width of 15 feet, what is its length?
5. How many square inches are in a square yard?

ANSWERS

1. 3 yards × 2 yards = 6 square yards
   6 × 9 = 54 square feet
2. 80 feet × 80 feet = 6,400 square feet
   6,400 × $100 = $640,000
3. 30 feet × 30 feet = 900 square feet
   100 square yards × $15 = $1,500
4. Area = length × width
   3,000 square feet = length × 15 feet
   200 feet = length
5. 12 × 12 × 9 = 1,296 square inches

3 Perimeters of Rectangles

A perimeter is a border or outer boundary. The perimeter of a rectangle is found by using the formula:

(2 × length) + (2 × width)

Problem 1:

How much is the perimeter of the following figure?

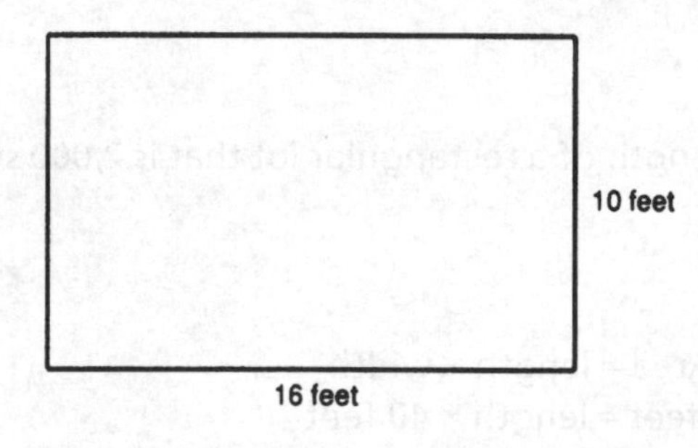

Solution:

Perimeter = (2 × length) + (2 × width)
= (2 × 16 feet) + (2 × 10 feet)
= 32 feet + 20 feet
= 52 feet

Problem 2:

What is the perimeter of a rectangular field if its length is 100 yards and its width is 30 yards?

Solution:

Perimeter = (2 × length) + (2 × width)
= (2 × 100 yards) + (2 × 30 yards)

$= 200 \text{ yards} + 60 \text{ yards}$
$= 260 \text{ yards}$

And now for something just a little bit different.

Problem 3:

What is the perimeter of a square whose side is 6 inches?

Solution:

$$\begin{aligned}\text{Perimeter} &= (2 \times \text{length}) + (2 \times \text{width})\\ &= (2 \times 6 \text{ inches}) + (2 \times 6 \text{ inches})\\ &= 12 \text{ inches} + 12 \text{ inches}\\ &= 24 \text{ inches}\end{aligned}$$

Of course, we could have taken a shortcut by just multiplying 6 inches by 4, since the length and width of squares are equal.

Problem 4:

How much would it cost to build a fence around a lot that is 200 feet long and 80 feet wide if fencing cost $8 a foot?

Solution:

$$\begin{aligned}\text{Perimeter} &= (2 \times \text{length}) + (2 \times \text{width})\\ &= (2 \times 200 \text{ feet}) + (2 \times 80 \text{ feet})\\ &= 400 \text{ feet} + 160 \text{ feet}\\ &= 560 \text{ feet}\\ \text{Cost} &= 560 \times \$8\\ &= \$4{,}480\end{aligned}$$

Problem 5:

How much would it cost to build a wall around the perimeter of a lot that is 60 feet long and 30 feet wide, if it cost $20 to build each foot of the wall?

Solution:

$$\begin{aligned}\text{Perimeter} &= (2 \times \text{length}) + (2 \times \text{width})\\ &= (2 \times 60 \text{ feet}) + (2 \times 30 \text{ feet})\\ &= 120 \text{ feet} + 60 \text{ feet}\\ &= 180 \text{ feet}\\ \text{Cost} &= 180 \text{ feet} \times \$20\\ &= \$3{,}600\end{aligned}$$

SELF-TEST 3

1. Find the perimeter of a field that is 500 feet long and 70 feet wide.
2. Find the perimeter of a square lot whose side is 40 feet.
3. How much would it cost to put a fence around a field that is 40 yards long and 20 yards wide if fencing cost $5 a foot (not a yard)?
4. How much would it cost to put fencing around a square lot whose side is 50 feet if fencing cost $8 a foot?

ANSWERS

1. Perimeter = (2 × length) + (2 × width)
   = (2 × 500 feet) + (2 × 70 feet)
   = 1,140 feet
2. 4 × 40 feet = 160 feet
3. (2 × 40 yards) + (2 × 20 yards) = 120 yards
   $5 a foot = $15 a yard
   120 × $15 = $1,800
4. 4 × 50 feet = 200 feet
   200 feet × $8 = $1,600

2 Areas of Triangles

The area of a triangle is one-half the base times the height. No ifs, ands, or buts. The great truth of this statement should become evident when we look at the following figure.

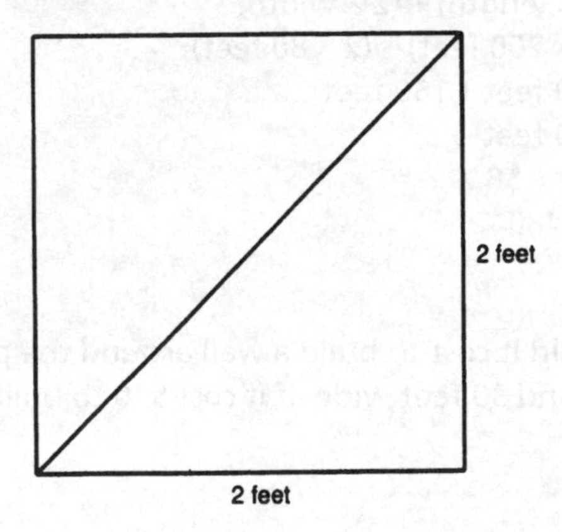

How much is the area of the square? Obviously, it's 4 square feet. Each triangle occupies half the area of the square. Now, how long is the base of the bottom triangle? It's 2 feet. And its height? Also 2 feet.

Problem 1:

Plug the numbers into the formula and see what you get for the area of the triangle:

$$\text{Area} = \frac{1}{2}\ \text{base} \times \text{height}$$

Solution:

$$\begin{aligned}\text{Area} &= \frac{1}{2}\ \text{base} \times \text{height}\\ &= \frac{1}{2} \times 2\ \text{feet} \times 2\ \text{feet}\\ &= \frac{1}{2} \times 4\ \text{square feet}\\ &= 2\ \text{square feet}\end{aligned}$$

Problem 2:

If a triangle had a base of 4 feet and a height of 6 feet, how much is its area?

Solution:

$$\begin{aligned}\text{Area} &= \frac{1}{2}\ \text{base} \times \text{height}\\ &= \frac{1}{2} \times 4\ \text{feet} \times 6\ \text{feet}\\ &= \frac{1}{2} \times 24\ \text{square feet}\\ &= 12\ \text{square feet}\end{aligned}$$

Problem 3:

Using the following figure, find the area of Triangle *ABC*.

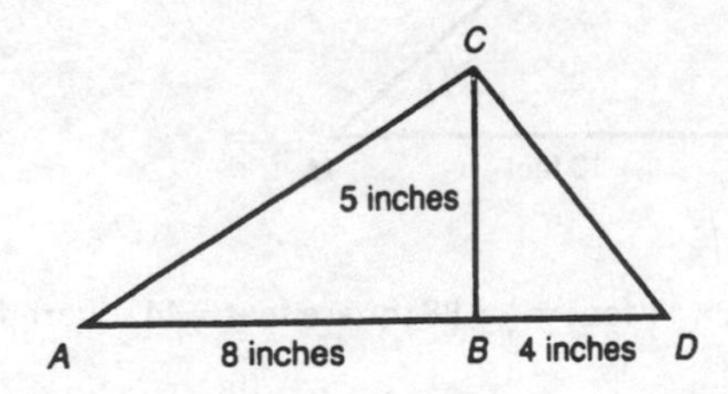

Solution:

$$\begin{aligned}\text{Area} &= \frac{1}{2}\ \text{base} \times \text{height}\\ &= \frac{1}{2} \times 8\ \text{inches} \times 5\ \text{inches}\end{aligned}$$

$= \frac{1}{2} \times 40$ square inches

$= 20$ square inches

Problem 4:

Using the same figure, find the area of Triangle *BCD*.

Solution:

Area $= \frac{1}{2}$ base $\times$ height

$= \frac{1}{2} \times 4$ inches $\times$ 5 inches

$= \frac{1}{2} \times 20$ square inches

$= 10$ square inches

SELF-TEST 4

1. Using the figure below, find the area of Triangle *EFG*.
2. Using the figure below, find the area of Triangle *FGH*.
3. If a triangle has a base of 15 inches and a height of 12 inches, how much is its area?

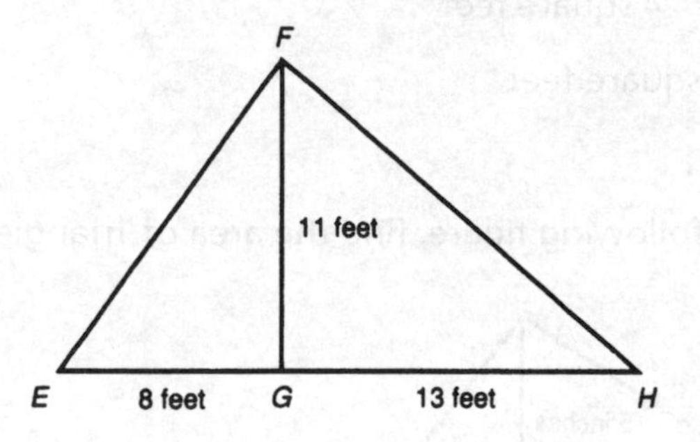

ANSWERS

1. Area $= \frac{1}{2} \times 8$ feet $\times$ 11 feet $= \frac{1}{2} \times 88$ square feet $= 44$ square feet
2. Area $= \frac{1}{2} \times 13$ feet $\times$ 11 feet $= \frac{1}{2} \times 143$ square feet $= 71.5$ square feet
3. Area $= \frac{1}{2} \times 15$ inches $\times$ 12 inches $= \frac{1}{2} \times 180$ square inches $= 90$ square inches

16 Finding the Circumferences and Areas of Circles

What if a UFO landed one night on a beach, abducted a few midnight swimmers, and then took off? Suppose it left an imprint in the sand that measured 1,000 feet in circumference. Just how big, then, was that UFO?

The circumference of a circle is the distance around that circle. So we know, then, that that UFO had a circular base that measured 1,000 feet around.

The next time a UFO with a circular base lands in *your* neighborhood, you'll be ready to measure it. Actually, there are many more down-to-earth reasons we can find to measure circles.

1 Circumferences of Circles

There are actually two ways for measuring the circumference of a circle. Besides walking around that circle, you can also walk directly across the circle. The path you'd walk is the diameter of the circle. So the diameter is a straight line, which passes through the center of the circle. The diameter is designated by the letter D, and is drawn in the circle that follows.

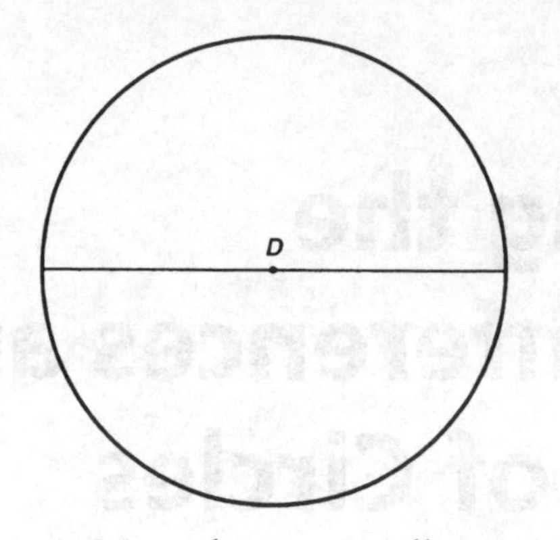

More than two millenniums ago, the Greeks found that the circumference of a circle was 3 1/7 times its diameter. You can check this out yourself by drawing a circle in the sand using a piece of string, and then using the string to measure the circumference of the circle.

The ratio of the circumference to the diameter, then, is 3 1/7:1, or 22/7:1. The Greeks designated the letter π (pronounced pie) to stand for 22/7. So the formula for the circumference of a circle is πD (or, $\pi \times D$).

Suppose you went for a walk around a circle. How far would you walk if the circle's diameter were 1 mile? You would walk 3 1/7 miles.

Problem 1:

If the diameter of a circle were 3 inches, what would the circumference of that circle be?

Solution:

$$\text{Circumference} = \pi \times D$$
$$= \frac{22}{7} \times \frac{3}{1} \text{ inches}$$
$$= \frac{66}{7} \text{ inches}$$
$$= 9\frac{3}{7} \text{ inches}$$

Problem 2:

Find the circumference of a circle whose diameter is 14 feet.

Solution:

$$\text{Circumference} = \pi \times D$$
$$= \frac{22}{7} \times \frac{14}{1} \text{ feet}$$

$$= \frac{22}{1} \times \frac{2}{1} \text{ feet}$$
$$= 44 \text{ feet}$$

Did you notice the little trick pulled off here? Since 7 goes into 14 twice, the 7 under the 22 cancelled out and the 14 over the 1 became a 2. This is in accord with the law of arithmetic that states that what you do to the top (divide by 7), you must also do to the bottom. (If you still don't understand how or why we divided 7 into 14, then please review frame 5 of chapter 8.)

Now we'll add a new wrinkle.

Problem 3:

If the circumference of a circle is 8 inches, how much is the diameter?

Solution:

$$\text{Circumference} = \pi \times D$$
$$8 \text{ inches} = \frac{22}{7} D$$
$$56 \text{ inches} = 22\,D$$
$$2.5 \text{ inches} = D$$

Problem 4:

If the circumference of a circle is 12 feet, how much is the diameter?

Solution:

$$\text{Circumference} = \pi \times D$$
$$12 \text{ feet} = \frac{22}{7} D$$
$$84 \text{ feet} = 22D$$
$$3.8 \text{ feet} = D$$

SELF-TEST 1

1. How much is the circumference of the circle in the figure here?

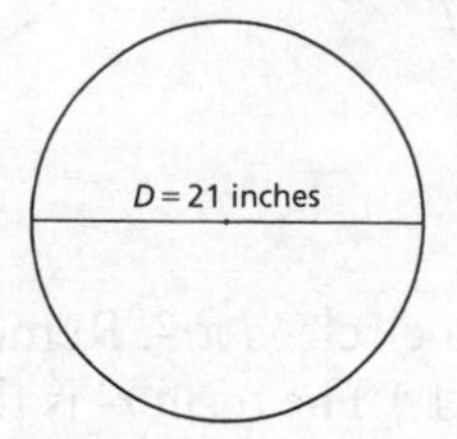

2. If the diameter of a circle is 9 feet, how much is its circumference?
3. If the circumference of a circle is 14 inches, how much is its diameter?
4. If the circumference of a circle is 9 feet, how much is its diameter?

ANSWERS

1. Circumference $= \frac{22}{\underset{1}{7}} \times \frac{\overset{3}{21}}{1}$ inches

   $= 66$ inches

2. Circumference $= \frac{22}{7} \times \frac{9}{1}$ feet

   $= \frac{198}{7}$ feet

   $= 28\frac{2}{7}$ feet

3. $14 \text{ inches} = \frac{22}{7}D$

   $98 \text{ inches} = 22D$
   $4.5 \text{ inches} = D$

4. $9 \text{ feet} = \frac{22}{7}D$

   $63 \text{ feet} = 22D$
   $2.9 \text{ feet} = D$

2 The Area of a Circle

To find the area of a circle, we need to introduce one new term, the radius, which is represented by r. Circle A has a radius of 4 inches, as shown.

The radius is any straight line that radiates from the center of the circle to some point on the circumference. By definition, all radii (the plural of radius) of the same circle are equal.

If the radius of a circle is 4 inches, how much is the diameter of that circle? It must be 8 inches, since the diameter is a straight line composed of two radii. Circle B shows two radii of 3 inches each, which equals a diameter of 6 inches.

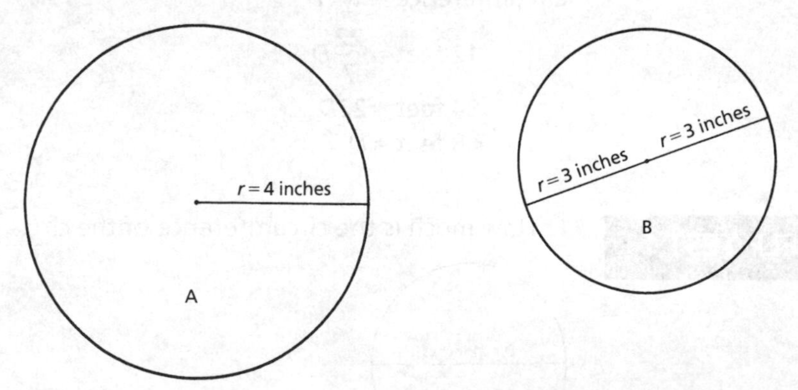

The formula for the area of a circle is πr^2. Remember squares? (If not, check back in chapter 13, frame 1.) The term r^2 is the same as saying $r \times r$.

Then why don't we just say $r \times r$? Because mathematicians like to save time. They save about three-tenths of a second by writing r^2 rather than $r \times r$. And they save even more time by writing r^4 rather than $r \times r \times r \times r$.

Now there are those who would challenge the validity of the formula for the area of a circle: πr^2 (pronounced "pie *r* square"). Once upon a time a faraway land was taken over by illiterates. Naturally, mathematicians were held in the greatest of scorn and had trouble getting any form of employment. One such mathematician, who had a Ph.D. from a famous university and had written many learned textbooks, applied for a job as a dishwasher in a restaurant. The owner asked him for his qualifications and seemed unimpressed even after the mathematician had given him his résumé, along with an imposing list of publications.

"Tell me a mathematical formula," said the restaurant owner.

"πr^2," replied the mathematician.

"I am sorry," said the owner. "You have given me an incorrect formula. Pie are *round; cake* are square."

Here are some problems that involve π and squares.

Problem 1:

If the radius of a circle is 6 units, what is its area?

Solution:

$$\begin{aligned} \text{Area} &= \pi r^2 \\ &= \frac{22}{7} \times \frac{6^2}{1} \\ &= \frac{22}{7} \times \frac{36}{1} \\ &= \frac{792}{7} \\ &= 113\frac{1}{7} \text{ square units} \end{aligned}$$

Problem 2:

What is the area of Circle B, shown earlier?

Solution:

$$\begin{aligned} \text{Area} &= \pi r^2 \\ &= \frac{22}{7} \times \frac{3 \text{ inches}^2}{1} \end{aligned}$$

$$= \frac{22}{7} \times \frac{9 \text{ square inches}}{1}$$

$$= \frac{198 \text{ square inches}}{7}$$

$$= 28\frac{2}{7} \text{ square inches}$$

Problem 3:

If the diameter of a circle is 10 units, what is its area?

Solution:

Area $= \pi r^2$

$$= \frac{22}{7} \times \frac{5^2}{1}$$

$$= \frac{22}{7} \times \frac{25}{1}$$

$$= \frac{550}{7}$$

$$= 78\frac{4}{7} \text{ square units}$$

Problem 4:

If the diameter of a circle is 8 feet, what is its area?

Solution:

Area $= \pi r^2$

$$= \frac{22}{7} \times \frac{4 \text{ feet}^2}{1}$$

$$= \frac{22}{7} \times \frac{16 \text{ square feet}}{1}$$

$$= \frac{352 \text{ square feet}}{7}$$

$$= 50\frac{2}{7} \text{ square feet}$$

SELF-TEST 2

1. If the radius of a circle is 4 inches, what is its area?
2. If the radius of a circle is 9 feet, what is its area?
3. If the diameter of a circle is 12 inches, what is its area?

4. If the diameter of a circle is 14 feet, what is its area?

ANSWERS

1. $\text{Area} = \pi r^2$

$$= \frac{22}{7} \times \frac{4 \text{ inches}^2}{1} = \frac{22}{7} \times \frac{16 \text{ square inches}}{1} = \frac{352 \text{ square inches}}{7}$$

$$= 50\frac{2}{7} \text{ square inches}$$

2. $\text{Area} = \pi r^2$

$$= \frac{22}{7} \times \frac{9 \text{ feet}^2}{1}$$

$$= \frac{22}{7} \times \frac{81 \text{ square feet}}{1} = \frac{1{,}782 \text{ square feet}}{7}$$

$$= 254\frac{4}{7} \text{ square feet}$$

3. $\text{Area} = \pi r^2$

$$= \frac{22}{7} \times \frac{6 \text{ inches}^2}{1}$$

$$= \frac{22}{7} \times \frac{36 \text{ square inches}}{1}$$

$$= \frac{792 \text{ square inches}}{7}$$

$$= 113\frac{1}{7} \text{ square inches}$$

4. $\text{Area} = \pi r^2$

$$= \frac{22}{7} \times \frac{7 \text{ feet}^2}{1}$$

$$= \frac{22}{7} \times \frac{49 \text{ square feet}}{1}$$

$$= \frac{22}{1} \times \frac{7 \text{ square feet}}{1}$$

$$= 154 \text{ square feet}$$

17 Rate, Time, and Distance Problems

Americans love to pack up the car and hit the open road. Or, if they're going more than a couple of hundred miles, they like to fly. This raises three basic questions: (1) how far? (2) how fast? and (3) how long? Here we'll cover how to answer these questions.

1 The Magic Formula

One of the greatest formulas of all time is distance = rate × time, or, among the in-crowd, $d = r \times t$. We'll use this formula and its two spinoffs, $r = d/t$ and $t = d/r$, to solve every problem in this chapter.

In frame 3, we use $d = r \times t$ to tell us how far we've gone. In frame 4, we learn to measure how fast we've gone by using $r = d/t$. Frame 5 helps us to measure how long the trip took from $t = d/r$. And finally, we do some mixing and matching—that is, we'll *mix* all three types of problems in one section and you'll *match* each one with the proper formula and solve it. Doesn't this sound like a lot of fun?

2 The Terms

We need to go over each of the terms, d (distance), t (time), and r (rate). Distance may be expressed in terms of inches, feet, yards, meters, miles, or kilometers. Don't worry, we'll be using miles virtually all of the time.

Rate, or the rate of speed, is usually stated in miles per hour. In the United States, we're so casual that we often don't bother to state "miles per hour," or even m.p.h. On our highways, the signs usually say, "Speed Limit: 60." In Canada, they're a bit more formal, expressing their limit of 100 kilometers per hour, and occasionally, they'll list the 62.5 m.p.h. equivalent below.

Time may be stated in seconds, minutes, hours, days, and so forth. Since we'll be dealing here with travel problems, the time element will generally be answered in terms of hours.

3 Finding Distance

We'll begin with an obvious problem. How far does a plane traveling at a speed of 500 m.p.h. go in 1 hour? It goes 500 miles. What distance does it go in 2 hours? Obviously, it goes 1,000 miles.

This gives us our formula: distance = rate × time. In this last instance, distance = 500 m.p.h. × 2 hours.

Problem 1:

A car averaging 55 m.p.h. reaches its destination in 3 1/2 hours. How far did it travel?

Solution:

$d = r \times t$
$= 55 \times 3.5$
$= 192.5$ miles

Since the problems we are doing involve rates expressed in miles per hour, we don't bother to write "m.p.h."; since time is expressed in hours, we don't bother to write "hours."

Problem 2:

Jessica leaves her house at 9 A.M. and walks steadily at a rate of 3 1/2 m.p.h. until noon. How far did she walk?

Solution:

$d = r \times t$
$= 3.5 \times 3$
$= 10.5$ miles

Problem 3:

Joshua leaves on a trip at 10 A.M. He stops 1 hour for lunch and arrives at his destination at 4:30 P.M. If he averaged 50 m.p.h., how far did he drive?

Solution:

$d = r \times t$
$= 50 \times 5.5$
$= 275$ miles

Problem 4:

Elizabeth started driving at 3 P.M., averaging 55 m.p.h. She stopped for 1 hour to have supper at 6 P.M. After supper she drove at 50 m.p.h. until 9:30 P.M. How far did she travel?

Solution:

We'll break down this problem into two parts: (1) the trip before supper and (2) the trip after supper.

(1) $d = r \times t$
$= 55 \times 3$
$= 165$ miles

(2) $d = r \times t$
$= 50 \times 2.5$
$= 125$ miles

$165 + 125 = 290$ miles

SELF-TEST 1

1. If a boat traveled for 6 1/2 hours at 25 m.p.h., how far would it travel?
2. If you left on a trip at 9:30 A.M. and drove steadily at 52 m.p.h. until 1:45 P.M., how far did you drive?
3. If a plane took off at 3 P.M. and flew at 450 m.p.h. until 4:30 P.M., and then took off again at 5:30 P.M. and few at 500 m.p.h. until 7:30 P.M., how far did it fly?
4. John left work at 5 P.M. and walked at 3 m.p.h. until 6:30 P.M, then he took a bus the rest of the way home. How far does John live from work if the bus traveled at 15 m.p.h. and left him off at his door at 7:00 P.M?

ANSWERS

1. $d = r \times t$
$= 25 \times 6.5$
$= 162.5$ miles

2. $d = r \times t$
$= 52 \times 4.25$
$= 221$ miles

3. a. $d = r \times t$
$= 450 \times 1.5$
$= 675$ miles

b. $d = r \times t$
$= 500 \times 2$
$= 1{,}000$ miles
$675 + 1{,}000 = 1{,}675$ miles

4. a. $d = r \times t$
$= 3 \times 1.5$
$= 4.5$ miles

b. $d = r \times t$
$= 15 \times .5$
$= 7.5$ miles
$4.5 + 7.5 = 12$ miles

4 Finding Rate

If we know how far someone traveled and we know how long the trip took, we can find that person's average rate of speed or m.p.h. All we need to do is plug the distance and time data into the formula, $r = d/t$.

Are you interested in the derivation of $r = d/t$? Well, I thought you'd never ask.

$$d = r \times t$$

$$\frac{d}{t} = \frac{r \times t}{t}$$

$$\frac{d}{t} = \frac{r \times \not{t}}{\not{t}}$$

$$\frac{d}{t} = r$$

or

$$r = \frac{d}{t}$$

Problem 1:

A group of cyclists left on a 100-mile trip at noon and arrived at their destination at 8 P.M. What was their average rate of speed (i.e., m.p.h.)?

Solution:

$$r = \frac{d}{t}$$

$$r = \frac{100}{8}$$

$$r = 12.5 \text{ m.p.h.}$$

Problem 2:

Cindy drove from Washington, D.C., to a suburb of Boston—a distance of 470 miles. She left Washington at 8 A.M. and arrived at her destination at 8 P.M. She stopped along the way for a total of 2 1/2 hours for meals and rest. What was her average rate of speed?

Solution:

$$r=\frac{d}{t}$$

$$r=\frac{470}{9.5}$$

$$=9.5\overline{)470}$$

$$
\begin{array}{r}
49.47 \\
=19\overline{)940.00} \\
\underline{-76}^{\text{x xx}} \\
180 \\
\underline{-171} \\
90 \\
\underline{-76} \\
140 \\
\underline{-133}
\end{array}
$$

$$=49.47=49.5 \text{ m.p.h}$$

To make our division simpler, we've doubled 9.5 to 19 (to remove the decimal) and then doubled 470 to 940. By doing this, we're dividing by a smaller number (19 instead of 95).

Problem 3:

Susan walked uphill from 9 A.M. until noon, covering a distance of 7 miles. At 1 P.M. she began her return trip, which took until 3 P.M. What was her average rate of speed?

Solution:

$$r=\frac{d}{t}$$

$$r=\frac{14}{5}$$

$$5\overline{)14.40}\quad 2.8$$

$r = 2.8$ m.p.h.

How about making some use of our algebra?

Problem 4:

A passenger train starts from New York and a freight train starts from San Francisco—the distance between the cities is 2,800 miles. As they head toward each other, the passenger train is moving at twice the rate of the speed of the freight train. After 10 hours they are still 1,600 miles apart. Find the rate of each train.

Solution:

Let r = the rate of the freight train
Let $2r$ = the rate of the passenger train

Since distance $= r \times t$, the distance traveled by the freight train is $r \times 10$, or $10r$. The distance traveled by the passenger train is $2r \times 10$, or $20r$.

The two trains covered a distance of $30r$ ($10r + 20r$). Since they covered a total of 1,200 miles between them (2,800 − 1,600), we can set:

$$30r = 1{,}200$$
$$r = 40$$
$$2r = 80$$

To use the $r = d/t$ formula to solve this problem, let r = the combined rate of the passenger and freight trains. We know that they covered a distance of 1,200 miles in 10 hours. So we can substitute these numbers into the formula:

$$r = \frac{d}{t}$$

$$r = \frac{1{,}200}{10}$$

$$r = 120$$

This leaves us with figuring out how fast the passenger and freight trains were traveling. So, let x = the rate of the freight train and $2x$ = the rate of the passenger train.

$$x + 2x = 120$$
$$3x = 120$$
$$x = 40$$
$$2x = 80$$

Which method is better? I like the first one better because it's shorter. But either method is okay because they both give us the right answers.

Problem 5:

Two cars started driving toward each other at 8 A.M. on the New York State Thruway. One started from Albany and the other from Buffalo, 300 miles away. One car drove 10 miles an hour faster than the other car. At 10 A.M., the cars were 80 miles apart. How fast were the cars going?

Solution:

Let r = the rate of the slower car
Let $r + 10$ = the rate of the faster car
Distance traveled by slower car in 2 hours = $2r$
Distance traveled by faster car in 2 hours = $2(r + 10) = 2r + 20$
Distance traveled by both cars = $2r + 2r + 20$
Distance traveled by both cars in 2 hours = 220 miles (300 − 80)

$$220 = 2r + 2r + 20$$
$$220 = 4r + 20$$
$$200 = 4r$$
$$50 = r$$
$$60 = r + 10$$

SELF-TEST 2

1. If a plane went 2,700 miles in 4 1/2 hours, what was its average rate of speed?
2. If Bob left his house at 7 A.M. and rode his bicycle 53 miles by 11 A.M., what was his average rate of speed?
3. Doreen ran 8 miles in 45 minutes. What was her average rate of speed?
4. If Joseph drove from home to work, a distance of 30 miles, in half an hour, and returned by a different route in 45 minutes, what was his average rate of speed?
5. Two planes begin to fly toward each other from points 8,000 miles apart, and one plane is flying one and a half times as fast as the

other. If after 3 hours they are still 5,000 miles apart, how fast is each plane flying?

6. Two trains start off from the same station and travel in opposite directions. After 4 hours they are 680 miles apart. If the first train is traveling at a rate of 20 m.p.h. faster than the second train, at what speeds are the trains traveling?

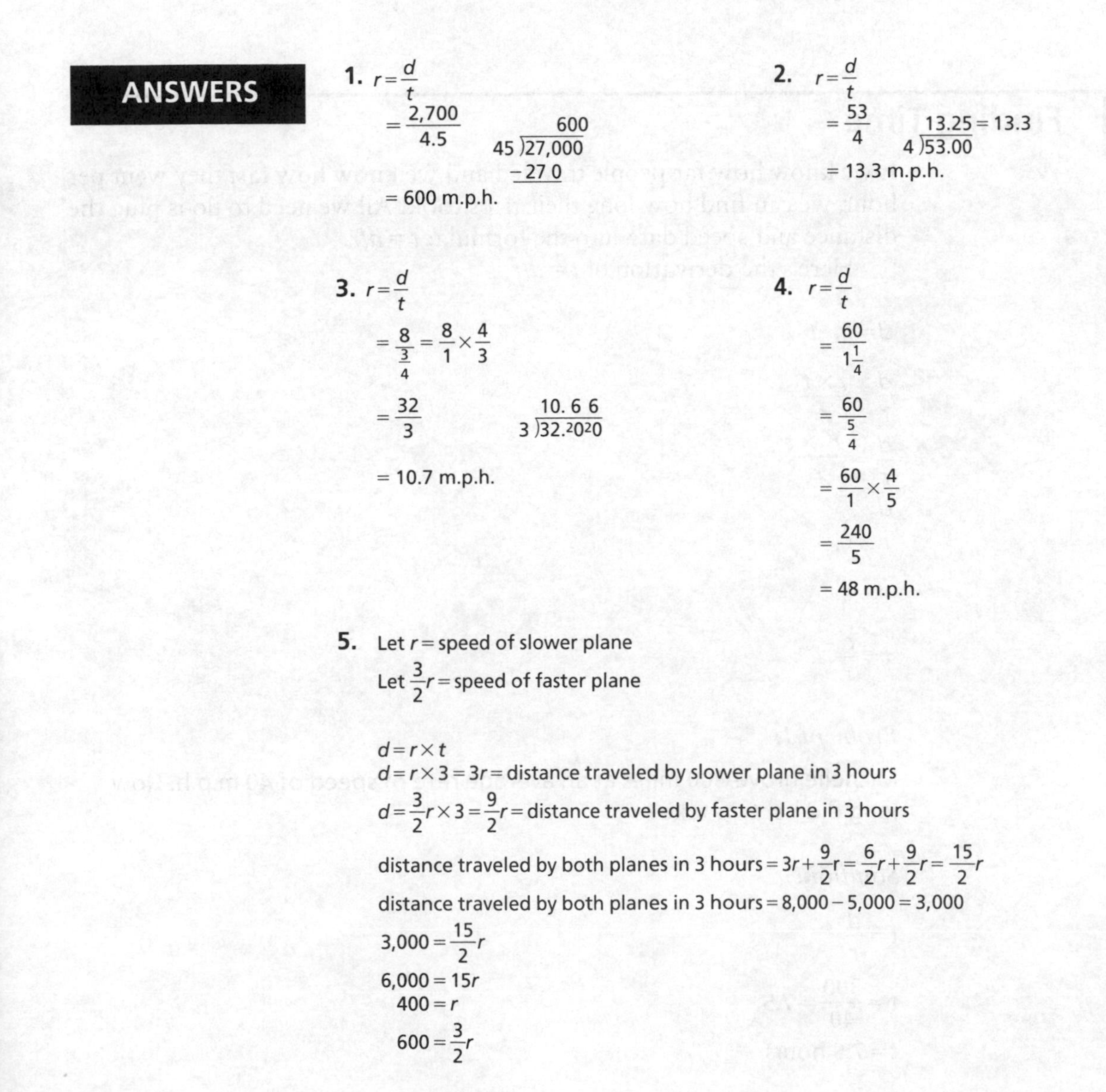

ANSWERS

1. $r=\frac{d}{t}$

$=\frac{2{,}700}{4.5}$

$$\begin{array}{r} 600 \\ 45\,\overline{)27{,}000} \\ \underline{-27\,0} \end{array}$$

$= 600$ m.p.h.

2. $r=\frac{d}{t}$

$=\frac{53}{4}$

$$\begin{array}{r} 13.25 = 13.3 \\ 4\,\overline{)53.00} \end{array}$$

$= 13.3$ m.p.h.

3. $r=\frac{d}{t}$

$=\frac{8}{\frac{3}{4}}=\frac{8}{1}\times\frac{4}{3}$

$=\frac{32}{3}$

$$\begin{array}{r} 10.6\,6 \\ 3\,\overline{)32.{}^{2}0{}^{2}0} \end{array}$$

$= 10.7$ m.p.h.

4. $r=\frac{d}{t}$

$=\frac{60}{1\frac{1}{4}}$

$=\frac{60}{\frac{5}{4}}$

$=\frac{60}{1}\times\frac{4}{5}$

$=\frac{240}{5}$

$= 48$ m.p.h.

5. Let $r=$ speed of slower plane

Let $\frac{3}{2}r=$ speed of faster plane

$d=r\times t$

$d=r\times 3=3r=$ distance traveled by slower plane in 3 hours

$d=\frac{3}{2}r\times 3=\frac{9}{2}r=$ distance traveled by faster plane in 3 hours

distance traveled by both planes in 3 hours $=3r+\frac{9}{2}r=\frac{6}{2}r+\frac{9}{2}r=\frac{15}{2}r$

distance traveled by both planes in 3 hours $=8{,}000-5{,}000=3{,}000$

$3{,}000=\frac{15}{2}r$

$6{,}000=15r$

$400=r$

$600=\frac{3}{2}r$

6. Let $r =$ speed of slower train
Let $r + 20 =$ speed of faster train
$d = r \times t$
$d = r \times 4 = 4r =$ distance traveled by slower train in 4 hours
$d = (r + 20)4 = 4r + 80 =$ distance traveled by faster train in 4 hours
distance traveled by both trains in 4 hours $= 4r + 4r + 80 = 8r + 80$
distance traveled by both trains in 4 hours $= 680$
$680 = 8r + 80$
$600 = 8r$
$75 = r$
$95 = r + 20$

5 Finding Time

If we know how far people traveled and we know how fast they went per hour, we can find how long their trips took. All we need to do is plug the distance and speed data into the formula: $t = d/r$.

Here's the derivation of $t = d/r$.

$$d = r \times t$$

$$\frac{d}{r} = \frac{r \times t}{r}$$

$$\frac{d}{r} = \frac{\not{r} \times t}{\not{r}}$$

$$\frac{d}{r} = t$$

or

$$t = \frac{d}{r}$$

Problem 1:

Michelle drove 300 miles at an average rate of speed of 40 m.p.h. How long did her trip take?

Solution:

$$t = \frac{d}{r}$$

$$t = \frac{300}{40} = 7.5$$

$t = 7.5$ hours

Problem 2:

Mark drove to work, a distance of 40 miles, at 50 m.p.h. He drove home from work at a speed of 45 m.p.h. How long did it take him to drive to and from work?

Solution:

We need to divide the problem into two parts: (1) the time it took to drive to work and (2) the time it took to drive home.

(1) $t = \dfrac{d}{r}$ (2) $t = \dfrac{d}{r}$

$t = \dfrac{40}{50} = \dfrac{4}{5}$ hours $t = \dfrac{40}{45} = \dfrac{8}{9}$ hours

$$(1) + (2) = \frac{4}{5} + \frac{8}{9} = \frac{4 \times 9}{5 \times 9} + \frac{8 \times 5}{9 \times 5} = \frac{36}{45} + \frac{40}{45} = \frac{76}{45} = 1\frac{31}{45} \text{ hours}$$

Now suppose we want to convert 1 31/45 hours to hours and minutes. To start, we'll express 1 31/45 in decimal form:

$1\dfrac{31}{45}$ hours = 1.6889 hours

```
        .688888 = .6889
  45 )31.000000
     -270xxxxx
        4 00
       -3 60
          400
         -360
           400
          -360
            400
           -360
             400
            -360
```

The .6889 represents part of an hour. Since there are 60 minutes in an hour, we can find the number of minutes in .6889 hours by multiplying .6889 by 60:

.6889 hours = .6889(60) = 41.334 minutes

Therefore, 1 31/45 hours = 1 hour 41.334 minutes.

Let's take this a step further and convert hours and minutes to hours, minutes, and seconds. The .334 minutes represents part of a minute.

Since there are 60 seconds in a minute, we can find the number of seconds in .334 minutes by multiplying .334 by 60:

.334 minutes $\times$ 60 = 20 seconds

So 1 31/45 hours = 1 hour, 41 minutes, and 20 seconds.

In summary, to convert a portion of an hour to minutes, multiply by 60 and change the units to minutes. To convert a portion of a minute to seconds, also multiply by 60 and change the units to seconds.

You don't need to carry out the solution this far—1 31/45 hours will suffice. I just wanted to show off a little (and to demonstrate that with the math you've had so far, you could show off a little too).

Problem 3:

Ms. Jones left home at 7 A.M. and rode her bicycle to work. She covered the distance of 8 miles at a speed of 12 m.p.h. She walked home at a speed of 4 m.p.h. How long did her trip to work and her trip home take her?

Solution:

This is a two-part question.

Trip to work:

$$t = \frac{d}{r}$$

$$t = \frac{8}{12}$$

$$= \frac{2}{3} \text{ hour}$$

Trip home:

$$t = \frac{d}{r}$$

$$t = \frac{8}{4}$$

$$= 2 \text{ hours}$$

Total commuting time: $2\frac{2}{3}$ hours or 2 hours 40 minutes

Problem 4:

Mr. and Mrs. Greenblatt left their home at 10 A.M., traveling in opposite directions. If Mrs. Greenblatt traveled at 48 m.p.h. and Mr. Greenblatt traveled at 44 m.p.h., at what time will they be 322 miles apart?

Solution:

Let's combine their two rates of speed:

$48 + 44 = 92$ m.p.h.

$$t = \frac{d}{r} = \frac{322}{92} = 3.5 \text{ hours}$$

$$\begin{array}{r} 3.5 \\ 92\overline{)322.0} \\ -276 \\ \hline 46\,0 \\ -46\,0 \\ \hline \end{array}$$

They will be 322 miles apart at 1:30 P.M.

SELF-TEST 3

1. Nicole walked at an average rate of speed of 3 1/2 m.p.h. and covered a distance of 14 miles. How long did she walk?
2. Harry ran to work, a distance of 10 miles, at 8 m.p.h. He took the bus home. If the bus traveled at a speed of 20 m.p.h., how long did it take Harry to get to and from work?
3. Two trains left the station traveling in opposite directions. One train was traveling at a rate of 70 m.p.h., and the other was traveling at the rate of 80 m.p.h. When they were 825 miles apart, for how long had they been traveling?
4. Alice got on Highway 80 at Omaha and headed west, averaging 55 m.p.h. John got on Highway 80 at the same time heading east from Omaha, traveling at an average rate of 52 m.p.h. When they were 535 miles apart, how long had they been traveling?

ANSWERS

1. $t = \frac{d}{r}$

   $t = \frac{14}{3.5} = 4$ hours

   $3.5\overline{)14} = 35\overline{)140} = 5\overline{)20} = 1\overline{)4}$ with quotient 4

   or $7\overline{)28}$ with quotient 4

2. (1) $t = \frac{d}{r} = \frac{10}{8} = 1\frac{1}{4}$ hours

   (2) $t = \frac{d}{r} = \frac{10}{20} = \frac{1}{2}$ hour

   $(1) + (2) = 1\frac{1}{4} + \frac{1}{2} = \frac{5}{4} + \frac{2}{4} = \frac{7}{4} = 1\frac{3}{4}$ hours

3.

$t = \frac{d}{r}$

r = combined rate for 1 hour

$= 80 + 70$

$= 150$

$t = \frac{825}{150}$

$= \frac{33}{6}$

$= \frac{11}{2}$

$= 5\frac{1}{2}$ hours

4.

$t = \frac{d}{r}$

r = combined rate for 1 hour

$= 55 + 52$

$= 107$

$t = \frac{535}{107}$

$= 5$ hours

6 Rate, Time, and Distance Problems

Nothing new is covered in this section. All we're going to do here is give you a self-test of all three types of problems. The trick is to figure out which type of problem you're doing and then to write down the proper formula and solve it.

If you're not confident about doing the $d = r \times t$ problems, you should reread frame 3 and retake Self-Test 1. If $r = d/t$ problems leave you at all uneasy, then you should reread frame 4 and retake Self-Test 2. And if you're not entirely clear on $t = d/r$ problems, then reread frame 5 and retake Self-Test 3.

SELF-TEST 4

1 If a bus traveling at an average speed of 46 m.p.h. reaches its destination in 2 1/2 hours, how far has it gone?

2. If a person needs 3 1/3 hours to walk 10 miles, how fast does she walk (i.e., in m.p.h.)?

3. If a plane flying at 500 m.p.h. flies 1,200 miles, how long does this trip take?

4. Max ran for 1 1/2 hours at a pace of 10 m.p.h. After resting, he walked back at a pace of 4 m.p.h. How long did his entire trip take?

5. Laura drove to work, a distance of 60 miles, covering the distance in 75 minutes. She took the bus home. This took 2 hours. What was her average rate of speed?

6. A plane flies at 550 m.p.h. for 45 minutes and 600 m.p.h. for 1 1/2 hours. How far did the plane fly?

7. Achilles and Hector decide to race each other around the world. They

start at noon from Troy. Achilles, heading due east, runs at an average speed of 8 m.p.h. Hector heads due west, running at 7.5 m.p.h. At what time will they be 93 miles apart?

8. Two trains start off from the same station and travel in opposite directions. After 6 hours they are 840 miles apart. If the first train is traveling at a rate of 20 m.p.h. faster than the second train, at what speeds are the trains traveling?

9. If a plane crashes on the border of the United States and Canada, where do they bury the survivors?

(Is there enough information here? Doesn't it depend on what country they're from? If that's what you think, I'll have to ask you to go back to frame 1. That's where you belong if you bury survivors.

Incidentally, you're a survivor. You've made it this far. Just another few chapters to go.)

ANSWERS

1. $d = r \times t$

$d = 46 \times \frac{5}{2}$

$= 23 \times 5$

$= 115$ miles

2. $r = \frac{d}{t}$

$r = \frac{10}{3\frac{1}{3}}$

$= \frac{10}{1} \div \frac{10}{3}$

$= \frac{10}{1} \times \frac{3}{10}$

$= 3$ m.p.h.

3. $t = \frac{d}{r}$

$t = \frac{1{,}200}{500}$

$= \frac{12}{5}$

$$5\overline{)12.^{2}0} = 2.4$$

$= 2.4$ hours or 2 hours 24 minutes

4. First part of trip: $= 1\frac{1}{2}$ hours

Second part of trip: $t = \frac{d}{r} = \frac{15}{4} = 3\frac{3}{4}$ hours

$1\frac{1}{2} + 3\frac{3}{4} = \frac{3}{2} + \frac{15}{4} = \frac{6}{4} + \frac{15}{4} = \frac{21}{4} = 5\frac{1}{4}$ hours

5. $r = \frac{d}{t}$

$r = \frac{120}{3\frac{1}{4}}$

$= \frac{120}{1} \div \frac{13}{4}$

$= \frac{120}{1} \times \frac{4}{13}$

$= \frac{480}{13}$

$r = 36.9$ m.p.h.

$$\begin{array}{r} 36.9 \\ 13\overline{)480.0} \\ -39 \\ \hline 90 \\ -78 \\ \hline 12\,0 \\ -11\,7 \\ \hline \end{array}$$

6. $d = r \times t$

a. $d = \frac{550}{1} \times \frac{3}{4}$

$= \frac{1{,}650}{4}$

$= 412.5$ miles

b. $d = \frac{600}{1} \times \frac{3}{2}$

$= \frac{1{,}800}{2}$

$= 900$ miles

$a + b = 412.5 + 900 = 1{,}312.5$ miles

7. $t = \frac{d}{r}$

Combined rate for Achilles and Hector: 15.5 m.p.h.

$t = \frac{93}{15.5}$ $\quad 15.5\overline{)93} = 155\overline{)930} = 31\overline{)186}$ (quotient 6; -186)

$t = 6$ hours

8. Let r = speed of slower train
Let $r + 20$ = speed of faster train
Distance covered by slower train in 6 hours $= r \times t = r \times 6$, or $6r$.
Distance covered by faster train in 6 hours $= r \times t = (r + 20)6 = 6r + 120$
Combined distance covered in 6 hours by both trains $= 6r + 6r + 120 = 12r + 120$
Given: Distance covered by both trains in 6 hours = 840 miles
$840 = 12r + 120$
$720 = 12r$
$60 = r$
$80 = r + 20$

7 Speed Limit Problems

Back in 1973, when the OPEC nations quadrupled the price of oil and gasoline shortages (or perhaps the fear of them) caused motorists to wait in line for hours at gas stations all around the country, the federal government passed a law. A national speed limit of 55 miles per hour was set. Not only would this conserve gasoline, since fuel economy falls sharply at speeds beyond 55 m.p.h., but this measure would also promote highway safety.

Both predictions came true. Fuel conservation was attained and highway deaths fell dramatically. But after a few years, a lot of people began to complain to Congress that at a speed of 55 m.p.h., it took too long to cover great distances. And furthermore, it was reasoned, there were plenty of stretches of highway in sparsely populated areas where speed limits could be safely raised.

Since the late 1980s, Congress has allowed the states to set their own limits in accordance with local driving conditions. This action gave rise to a whole new set of rate × time = distance problems. Here they come.

Problem 1:

How much time would you save by driving for 100 miles at 60 m.p.h. rather than 55 m.p.h.?

Solution:

$t = \frac{d}{r}$

at 55: $t = \frac{100}{55} = \frac{20}{11}$

$$11\overline{)20.000} = 1.818 = 1.82 \text{ hours}$$

$$\begin{array}{r} 20.000 \\ -11 \\ \hline 90 \\ -88 \\ \hline 20 \\ -11 \\ \hline 90 \\ -88 \\ \hline \end{array}$$

at 60: $t = \frac{100}{60} = \frac{10}{6} = \frac{5}{3}$ $\quad 3\overline{)5^{2}.000} = 1.666 = 1.67$ hours

time saved = 1.82 hours − 1.67 hours = .15 hour or 9 minutes

Problem 2:

How much time would you save if you drove 20 miles at 60 m.p.h. instead of at 50 m.p.h.?

Solution:

$t = \frac{d}{r}$

at 50: $t = \frac{20}{50}$

$= \frac{2}{5}$ hour = 24 minutes

at 60: $t = \frac{20}{60}$

$= \frac{1}{3}$ hour = 20 minutes

time saved = 4 minutes

Problem 3:

How much farther would you get if you drove for half an hour at 75 m.p.h. rather than 55 m.p.h.?

Solution:

$d = r \times t$

at 55: $d = 55 \times \frac{1}{2}$

$= 27\frac{1}{2}$ miles

at 75: $d = 75 \times \frac{1}{2}$

$= 37\frac{1}{2}$ miles

$37\frac{1}{2} - 27\frac{1}{2} = 10$ miles

One last question. This will appear to be a trick question, but if you put the right numbers into the right formula, you'll get it right.

Problem 4:

How much time would you save if you drove one mile at 65 m.p.h. rather than 55 m.p.h.?

Solution:

$t = \frac{d}{r}$

at 55: $t = \frac{1}{55}$ hour $= 55 \overline{)1.00000}$

$$\begin{array}{r} .01818 \\ 55\overline{)1.00000} \\ -55^{xxx} \\ \hline 450 \\ -440 \\ \hline 100 \\ -55 \\ \hline 450 \\ -440 \\ \hline \end{array}$$

at 65: $t = \frac{1}{65}$ hour $= 65 \overline{)1.00000}$

$$\begin{array}{r} .01538 \\ 65\overline{)1.00000} \\ -65^{xxx} \\ \hline 350 \\ -325 \\ \hline 250 \\ -195 \\ \hline 550 \\ -520 \\ \hline 30 \end{array}$$

Time saved:

$$\begin{array}{r} .01818 \text{ hour} \\ -.01538 \\ \hline .00280 \end{array}$$

How many minutes do you save?

$.0028 \times 60 = .168$ minutes

And how many seconds does that come to?

$.168 \times 60 = 10.08$ seconds

If you didn't carry this out as far as I did, or even if you didn't get it right, I wouldn't lose any sleep over it. Did you get at least two out of the first three problems right? Then go to Self-Test 5. If you didn't, please go back to frame 1 of this chapter. Remember, it's always better the second time around.

SELF-TEST 5

1. How much time would you save by driving for 250 miles at 60 m.p.h. rather than 55 m.p.h.?
2. How much time would you save if you drove 55 miles at 60 m.p.h. rather than at 55 m.p.h.?
3. How much time would you save if you drove for 10 miles at 75 m.p.h. rather than at 55 m.p.h.?
4. How much farther would you get if you drove for 10 minutes at 75 m.p.h. rather than at 55 m.p.h.?
5. How much farther would you get if you drove for 20 minutes at 60 m.p.h. rather than at 55 m.p.h.?

ANSWERS

1. $t = \frac{d}{r}$

at 55: $t = \frac{250}{55} = \frac{50}{11}$

$$11\overline{)50.000} = 4.545 = 4.545 \text{ hours}$$

$$\begin{array}{r} -44 \\ \hline 60 \\ -55 \\ \hline 50 \\ -44 \\ \hline 60 \\ -55 \\ \hline \end{array}$$

at 60: $t = \frac{250}{60} = \frac{25}{6}$ $\quad 6\overline{)25.0000} = 4.1666 = 4.167$ hours

$t = 4.545 - 4.167$
$= 0.3784$ hour
$t = 22.68$ minutes

$$\begin{array}{r} .378 \\ \times 60 \text{ minutes} \\ \hline 22.680 \text{ minutes} \end{array}$$

2. $t = \frac{d}{r}$

at 55: $t = \frac{55}{55} = 1$ hour

at 60: $t = \frac{55}{60} = \frac{11}{12}$

$$12\overline{)11.000} = .91666 = .917 \text{ hours}$$

$$\begin{array}{r} -10\,8 \\ \hline 20 \\ -12 \\ \hline 80 \\ -72 \\ \hline 8 \end{array}$$

$t = 1 - .917$
$= .083$ hour
$t = 4.98$ minutes

$$\begin{array}{r} .083 \\ \times 60 \text{ minutes} \\ \hline 4.980 \text{ minutes} \end{array}$$

3. $t = \frac{d}{r}$

at 55: $t = \frac{10}{55} = \frac{2}{11}$

```
      .1818 = .1818 hour
11 )2.0000
   -1 1 xxx
      90
     -88
      20
     -11
      90
     -88
```

at 75: $t = \frac{10}{75} = \frac{2}{15}$

```
      .1333 = .13 hour
15 )2.0000
   -1 5 xxx
      50
     -45
      50
     -45
      50
     -45
```

$t = .1818 - .1333$
$= .0485$ hour
$t = 2.91$ minutes, or 2 minutes 55 seconds

```
  .0485 hour
  ×60 minutes
2.9100 minutes
```

```
  .91
  ×60
 54.6
```

4. $d = r \times t$

at 55: $d = 55 \times \frac{1}{6} = \frac{55}{6} = 9\frac{1}{6}$ miles

at 75: $d = 75 \times \frac{1}{6} = \frac{75}{6} = 12\frac{1}{2}$ miles

$12\frac{1}{2} - 9\frac{1}{6} =$

$12\frac{3}{6} - 9\frac{1}{6} = 3\frac{2}{6} = 3\frac{1}{3}$

$d = 3\frac{1}{3}$ miles

5. $d = r \times t$

at 55: $d = 55 \times \frac{1}{3} = 18\frac{1}{3}$ miles

at 60: $d = 60 \times \frac{1}{3} = 20$ miles

$d = 20 - 18\frac{1}{3} = 1\frac{2}{3}$ miles

18 Finding x: Advanced Problems

In this chapter we'll set out to find the ages of various people, the identities of groups of numbers, and the composition of various mixtures of nuts. What do these different problems have in common? Each involves the search for x, the unknown.

These exercises will sharpen your algebraic reasoning, possibly get you a job at a carnival guessing people's ages, and qualify you for employment in a nut store. I know you can hardly wait to get started.

1 Age Problems

Let's go back to x, the great unknown. Surely there is some x you've always been curious about—perhaps it's somebody's age. Suppose someone said, "I'm 10 years older than George." If you knew that the person telling you this were 50, it wouldn't be hard to figure out that George is 40. Or, if this person, who you know is 50, said, "I'm twice as old as George," you'd immediately figure out that George is 25.

Believe it or not, you were using algebra to figure out George's age in both instances. If x were George's age and someone who was 50 said he was 10 years older than George, you would have a ready-made equation:

$x + 10 = 50$
$x = 40$

Let's take this one step further.

Problem 1:

If George's age were still x, the unknown, and someone who is 50 told you he was twice as old as George, write an equation to find George's age.

Solution:

$2x = 50$
$x = 25$

Those were *easy* ones. We'll try something a little harder.

Problem 2:

If I am 28, and I am 5 years less than three times your age, how old are you? How about, too young to figure this out? No, you wouldn't be. Remember to let x be the unknown (which is your age).

Solution:

$3x - 5 = 28$
$3x = 33$
$x = 11$

I hope you're having fun. Here's another one.

Problem 3:

A mother is twice as old as her daughter. Their combined ages equal 60. How old is the mother and how old is the daughter? Hint: The key here is to figure out what to have x represent.

Solution:

Let daughter's age $= x$
Let mother's age $= 2x$
$x + 2x = 60$
$3x = 60$
$x = 20$
$2x = 40$

Here's another one.

Problem 4:

Alice is twice as old as Jesse. Jesse is twice Marie's age. Their combined ages equal 105. How old is each?

Solution:

Let x = Marie's age
Let $2x$ = Jesse's age
Let $4x$ = Alice's age

$$x + 2x + 4x = 105$$
$$7x = 105$$
$$x = 15$$
$$2x = 30$$
$$4x = 60$$

How are you doing? If you need help in setting up equations, go to frame 1 of chapter 12, reread the entire chapter, and then begin this chapter again.

This next one is really pushing it. If you get it right, then I want you to pat yourself on the back for a job well done. If you've gotten everything in this section right up to now and you blow this one, don't worry about it. You can't win 'em all.

Problem 5:

Three years ago a mother was five times her daughter's age. Three years from now, she will be only three times as old as her daughter. How old is the daughter today? (Hint: Once you let x equal the daughter's age today, represent the mother's age 3 years ago and the mother's age 3 years from now.)

Solution:

Let x = daughter's age today
Let $x - 3$ = daughter's age 3 years ago
Let $5(x - 3)$ = mother's age 3 years ago
Let $5(x - 3) + 6$ = mother's age 3 years from now
Let $3(x + 3)$ = mother's age 3 years from now also
Therefore:

$$5(x - 3) + 6 = 3(x + 3)$$
$$5x - 15 + 6 = 3x + 9$$
$$5x - 9 = 3x + 9$$
$$5x = 3x + 18$$

$$2x = 18$$
$$x = 9$$

SELF-TEST 1

1. If Marcia is 4 years short of being twice Linda's age, and the sum of their ages is 107, how old are these women?
2. Ken is three times Janice's age and twice Bill's age. If their combined ages equal 55, how old are Ken, Janice, and Bill?
3. Alan is three times Sheila's age. In 5 years he'll be just twice her age. How old are Alan and Sheila?
4. Two years ago, Fred was four times Steve's age. Three years from now he will be just three times Steve's age. How old are Fred and Steve today?

ANSWERS

1. Let $x =$ Linda's age
   Let $2x - 4 =$ Marcia's age
   $x + 2x - 4 = 107$
   $3x - 4 = 107$
   $3x = 111$
   $x = 37$
   $2x - 4 = 70$

2. Let $x =$ Janice's age
   Let $3x =$ Ken's age
   Let $3/2x =$ Bill's age
   $x + 3x + \frac{3}{2}x = 55$
   $4x + \frac{3x}{2} = 55$
   $\frac{8x}{2} + \frac{3x}{2} = 55$
   $\frac{11x}{2} = 55$
   $11x = 110$
   $x = 10$
   $3x = 30$
   $\frac{3}{2x} = 15$

3. Let $x =$ Sheila's age
   Let $3x =$ Alan's age
   $3x + 5 = 2(x + 5)$
   $3x + 5 = 2x + 10$
   $3x = 2x + 5$
   $x = 5$
   $3x = 15$

4. Let $x =$ Steve's age now
   Let $4(x - 2) =$ Fred's age 2 years ago
   Let $4(x - 2) + 5 =$ Fred's age 3 years from now
   Let $3(x + 3) =$ Fred's age 3 years from now also
   $4(x - 2) + 5 = 3(x + 3)$
   $4x - 8 + 5 = 3x + 9$
   $4x - 3 = 3x + 9$
   $4x = 3x + 12$
   $x = 12$
   Fred is 42.

2 Finding the Numbers

After the workout you had in the last section, this one will be child's play. What we'll be doing here is finding the mystery numbers.

Problem 1:

Three consecutive numbers add up to 24. Find the numbers.

Solution:

Let x be the first number
Let $x+1$ be the second numbers
Let $x+2$ be the third number

$$x+x+1+x+2=24$$
$$3x+3=24$$
$$3x=21$$
$$x=7$$
$$x+1=8$$
$$x+2=9$$

Problem 2:

Four consecutive numbers add up to 82. Find the numbers.

Solution:

Let x be the first number
Let $x+1$ be the second number
Let $x+2$ be the third number
Let $x+3$ be the fourth number

$$x+x+1+x+2+x+3=82$$
$$4x+6=82$$
$$4x=76$$
$$x=19$$
$$x+1=20$$
$$x+2=21$$
$$x+3=22$$

Pretty easy, huh? Okay, we'll start to tighten the screws.

Problem 3:

Find three consecutive numbers the sum of which is -39.

Solution:

Let $x=$ the first number
Let $x+1=$ the second number
Let $x+2=$ the third number

$$x+x+1+x+2=-39$$

$$3x+3=-39$$
$$3x=-42$$
$$x=-14$$
$$x+1=-13$$
$$x+2=-12$$

Problem 4:

Find four consecutive numbers the sum of which is -2.

Solution:

Let $x=$ the first number
Let $x+1=$ the second number
Let $x+2=$ the third number
Let $x+3=$ the fourth number

$$x+x+1+x+2+x+3=-2$$
$$4x+6=-2$$
$$4x=-8$$
$$x=-2$$
$$x+1=-1$$
$$x+2=0$$
$$x+3=+1$$

Problem 5:

The sum of three numbers is -16. The second number is four times the first. The third number is 2 more than the second.

Solution:

Let $x=$ the first number
Let $4x=$ the second number
Let $4x+2=$ the third number

$$x+4x+4x+2=-16$$
$$9x+2=-16$$
$$9x=-18$$
$$x=-2$$
$$4x=-8$$
$$4x+2=-6$$

SELF-TEST 2

1. Find three consecutive numbers that add up to 42.
2. Find three consecutive numbers that add up to -3.
3. Three numbers add up to 82. The second number is four times as large as the first, and the third is 8 less than the second. Find the numbers.

4. The sum of three numbers is 5. The second number is three times the first. The third number is 10 less than the first. Find the numbers.

5. The sum of four numbers is 30. If the second number is twice as large as the first, the third is 2 less than the second, and the fourth is 3 larger than twice the third, find the numbers. (Hint: These are not round numbers.)

ANSWERS

1. Let x = the first number
Let $x+1$ = the second number
Let $x+2$ = the third number

$$x+x+1+x+2=42$$
$$3x+3=42$$
$$3x=39$$
$$x=13$$
$$x+1=14$$
$$x+2=15$$

2. Let x = the first number
Let $x+1$ = the second number
Let $x+2$ = the third number

$$x+x+1+x+2=-3$$
$$3x+3=-3$$
$$3x=-6$$
$$x=-2$$
$$x+1=-1$$
$$x+2=0$$

3. Let x = the first number
Let $4x$ = the second number
Let $4x-8$ = the third number

$$x+4x+4x-8=82$$
$$9x-8=82$$
$$9x=90$$
$$x=10$$
$$4x=40$$
$$4x-8=32$$

4. Let x = the first number
Let $3x$ = the second number
Let $x-10$ = the third number

$$x+3x+x-10=5$$
$$5x-10=5$$
$$5x=15$$
$$x=3$$
$$3x=9$$
$$x-10=-7$$

5. Let x = the first number
Let $2x$ = the second number
Let $2x-2$ = the third number
Let $2(2x-2)+3$ = the fourth number

$$x+2x+2x-2+2(2x-2)+3=30$$
$$x+2x+2x-2+4x-4+3=30$$
$$9x-3=30$$
$$9x=33$$
$$3x=11$$
$$x=3\frac{2}{3}$$
$$2x=7\frac{1}{3}$$
$$2x-2=5\frac{1}{3}$$
$$2(2x-2)+3=13\frac{2}{3}$$

3 Nut Problems

You know those cans of mixed nuts that are always about 70 percent peanuts? Well, suppose that we make up our own mixtures and go very easy on the peanuts.

Problem 1:

Let's make up a mixture of 10 pounds of nuts. We'll use some peanuts, which sell for \$2 per pound, and some almonds, which sell for \$4 per pound. If the nut mixture sells for \$3.50 per pound, how many pounds of peanuts and how many pounds of almonds did we use? Try to work it out. Use your algebra. To get started, let x = something (actually, pounds of something).

Solution:

Let x = pounds of peanuts
Let $10 - x$ = pounds of almonds

Why do we set things up this way? Because we need to find two things: the number of pounds of peanuts and the number of pounds of almonds used in the mixture.

Next step:

What is the value of the peanuts that go into the mixture? Since peanuts are \$2 per pound and we're using x pounds, their value is $\$2x$.

What is the value of the almonds?

$10 - x$ pounds at $\$4 = \$4(10 - x)$, or $\$40 - \$4x$

Next step:

What is the value of the whole mixture? Just add the value of the peanuts to the value of the almonds:

$\$2x + \$40 - \$4x$

There's just one missing ingredient. Since we already know that the mixture will sell for \$3.50 per pound and weighs 10 pounds, the value of the mixture is $10 \times \$3.50$, or \$35.

So, we can set up this equation:

$$\begin{aligned} \$2x + \$40 - \$4x &= \$35 \\ -\$2x + \$40 &= \$35 \\ -\$2x &= -\$5 \\ \$2x &= \$5 \\ 2x &= 5 \\ x &= 2\tfrac{1}{2} \\ 10 - x &= 7\tfrac{1}{2} \end{aligned}$$

I know that cancelling out those dollar signs is not the most elegant procedure, but it does lead us to our answer.

If you're puzzled by our going from $-\$2x = -\$5$ to $\$2x = \$5$, here's what we did. We multiplied both sides by -1. Remember, you can do virtually anything to one side of an equation if you do it to the other side as well.

Now we both know that I did all the work on this last one, so I'll let you do the work on the next problem.

Problem 2:

We're going to upgrade our mixture. Let's mix cashews ($5 per pound) with pecans ($8 per pound). We'll do a 50-pound mixture that will sell for $6.50 per pound.

Solution:

Let x = pounds of cashews
Let $50 - x$ = pounds of pecans
The value of the cashews is $5x. The value of the pecans is $\$8(50 - x)$, or $\$400 - \$8x$. And the value of the mixture is $50 \times \$6.50$, or $325.

$$\$5x + \$400 - \$8x = \$325$$
$$-\$3x + \$400 = \$325$$
$$-\$3x = -\$75$$
$$\$3x = \$75$$
$$3x = 75$$
$$x = 25$$
$$50 - x = 25$$

Notice that the price of the mixture ($6.50) split the difference between the cashews ($5) and the pecans ($8). So you could have saved yourself the trouble of all that work by just saying, since the price was split down the middle, that the mixture must be half cashews and half pecans. As you get used to working with numbers, this type of observation will become second nature.

These problems are so long that we'll go directly to the self-test. However, if you're not sure how to do them, please go back to frame 3 and go over the problems again.

SELF-TEST 3

1. A 20-pound mixture of peanuts and Brazil nuts sells for $6 a pound. If peanuts sell for $3 a pound and Brazil nuts for $7 a pound, how many pounds of peanuts and how many pounds of Brazil nuts are used in the mixture?

2. An 80-pound mixture of shelled walnuts and cashews is sold for $4 a pound. If the walnuts sell for $3 a pound and the cashews for $6 a pound, how many pounds of walnuts and how many pounds of cashews are in the mixture?

3. A 60-pound mixture of peanuts and shelled pistachio nuts sells for $5 a pound. If the peanuts sell for $2 a pound and the pistachios are $8 a pound, how many pounds of pistachios and how many pounds of peanuts are in the mixture?

ANSWERS

1. Let $x =$ pounds of peanuts
   Let $20 - x =$ pounds of Brazil nuts
   The value of peanuts in the mixture is $3*x*. The value of Brazil nuts in the mixture is $7(20 − *x*), or $140 − $7*x*. The entire mixture sells for $6 × 20, or $120.
   $$\$3x + \$140 - \$7x = \$120$$
   $$-\$4x + \$140 = \$120$$
   $$-\$4x = -\$20$$
   $$\$4x = \$20$$
   $$4x = 20$$
   $$x = 5$$
   $$20 - x = 15$$

2. Let $x =$ pounds of walnuts
   Let $80 - x =$ pounds of cashews
   The value of walnuts in the mixture is $3*x*. The value of cashews in the mixture is $6(80 − *x*), or $480 − $6*x*. The entire mixture sells for $4 × 80, or $320.
   $$\$3x + \$480 - \$6x = \$320$$
   $$-\$3x + \$480 = \$320$$
   $$-\$3x = -\$160$$
   $$\$3x = \$160$$
   $$3x = 160$$
   $$x = 53\frac{1}{3}$$
   $$80 - x = 26\frac{2}{3}$$

3. The easy way is to split the difference: 30 pounds of peanuts and 30 pounds of pistachio nuts. We can do this because the price of the mixture is $5, which is halfway between $2 and $8.
   Here's the entire solution for those who worked it out.
   Let $x =$ pounds of peanuts
   Let $60 - x =$ pounds of pistachios
   The value of the peanuts is $2*x*. The value of the pistachios is $8(60 − *x*), or $480 − $8*x*. The entire mixture sells for $5 × 60, or $300.
   $$\$2x + \$480 - \$8x = \$300$$
   $$-\$6x + \$480 = \$300$$
   $$-\$6x = -\$180$$
   $$\$6x = \$180$$
   $$6x = 180$$
   $$x = 30$$
   $$60 - x = 30$$

19 Interest Rates

When you borrow money, the interest rate you pay may be a lot higher than you bargained for. And when you receive interest for the money that you lend out, you're not necessarily receiving the rate of interest you think you are. All of which just goes to show that interest is not such a simple concept. But as we go from the simple interest rate to the compound and beyond, we will eventually end up with the true rate of interest.

1 Simple Interest

There is interest and there is *interest*. What's the difference? The difference can be *mucho dinero*, which, translated loosely from the original Spanish, means big bucks.

The great financier, Rothschild, was said to have lived on the interest of his interest. Or was it the interest of the interest of his interest? At any rate, the man was heavily into compound interest, which is the subject of the next section, and that is big bucks.

And simple interest? Well, simple interest is okay, but it's not all that lucrative. We'll illustrate with a simple problem. If you lent \$100 for 1 year to a friend, and she paid you 5% interest, how much money would she give you when she repaid the loan?

The answer is \$105: the \$100 principal of the loan plus \$5 interest. This \$5 is 5% of \$100.

How much money would she have paid you if the loan had been for 2 years? She would have paid you $110: the $100 principal, plus $10 interest ($5 for each year).

What if she had borrowed $100 for 10 years? Then, using the same 5% rate of simple interest, she would have paid you $150, or $100 principal plus $50 interest ($5 for each of the 10 years that the loan was outstanding).

Now we'll do simple interest loans for parts of a year.

Problem 1:

How much interest is paid on a $200 loan that is made for 6 months at an annual simple interest rate of 6%?

Solution:

Six percent of $200 is $12. Since the money was lent for just 6 months, or one-half of a year, the interest paid would be $6.

Problem 2:

How much interest is paid on a $1,000 loan that is made for 3 months at an annual simple interest rate of 8%?

Solution:

Eight percent of $1,000 is $80. Since the loan was made for just one-quarter of a year, $20 interest is paid.

It should be clear from these examples that the annual interest rate indicates the interest that would be paid if the loan were extended for 1 year. If the loan is for just 6 months, we can find the interest paid by multiplying the principal by one-half the annual interest rate. We did that in the first example: a $200 loan for 6 months at 6%. A fast way of figuring the interest would be to multiply $200 by 3%.

In the last problem we did—$1,000 for 3 months at an annual rate of 8%—we can calculate $1,000 at 2%. Not only does this shortcut save you time, it makes you look smart when you apply for a 90-day bank loan.

SELF-TEST 1

Calculate the simple interest paid for each of these problems:

1. $5,000 for 6 months at an annual rate of 8%.
2. $3,000 for 3 years at an annual rate of 7%.
3. $4,000 for 1 year at an annual rate of 9%.

4. $2,000 for one quarter at an annual rate of 12%.

ANSWERS

1. $\$5{,}000 \times .04 = \$200$
2. $\$3{,}000 \times 3(.07) = \$3{,}000 \times .21 = \$630$
3. $\$4{,}000 \times .09 = \$360$
4. $\$2{,}000 \times .03 = \$60$

2 Compound Interest

Back in the good old days during the Spanish-American War when I was growing up, banks used to pay interest quarterly. This was a practice that they kept up into the 1950s. On the first banking day of every quarter, a long line of old folks would be waiting to have the interest recorded in their savings passbooks. It wasn't really necessary for them to do this—the banks automatically credited their accounts—but you know how some old folks are.

Suppose you withdrew some of your savings a few days before the quarter ended. The banks could have made up signs—"Tough, we won't pay you a cent!" The money had to be there until the last day of the quarter for interest to be earned.

Today banks are a lot nicer—and, incidentally, competition among banks for deposits is a lot tougher. Now they'll pay interest, as they phrase it, "from day of deposit to day of withdrawal." And better yet, the interest they pay is compounded not annually, not quarterly, not monthly, not weekly, but daily. In other words, like Rothschild, even your interest is earning interest. But don't count on living off it.

Welcome to the magic world of compound interest. And welcome to the world of compound interest problems.

There are three ways to calculate compound interest: (1) use a table (these appear in various math texts—not this one, however—and there are books of compound interest tables published); (2) use a calculator; and (3) work it out with pencil and paper. Guess which method we're going to use?

To make this less intimidating, we'll make two concessions. Instead of compounding our interest daily, we'll do it quarterly. And we'll limit all of our problems to no more than 1 year.

Now hang on to your seat because I'm going to give you a very powerful formula that is used to calculate compound interest. Here it comes:

$$A = P(1 + r)^t$$

A stands for the amount of money that the lender ends up with after being repaid the initial principal, *P*, plus interest earned. The "1" is simply the number one. The letter r is the interest rate (for the time period covered). And t stands for the number of time periods.

If the time period is 1 year (i.e., interest is compounded annually and the loan is for 1 year), then t would be equal to 1. And the r would be the annual rate of interest. But interest may be compounded more often than annually. What if interest is compounded quarterly? If the annual rate of interest were 20% and the loan were for 1 year, how much would r be, and how much would t be?

The r would be 5% (one-quarter of 20%), and t would be 4 (interest is compounded four times in 1 year).

Just for the fun of it, suppose the annual rate of interest were 12% and interest were paid for 1 year compounded monthly. How much would r be, and how much would t be?

The r would be 1% (1/12 of 12), and the t would be 12 (interest is compounded 12 times a year).

I know all this is a lot of fun, but it won't do us much good unless we apply it to some problems. We'll start with a real easy one.

Problem 1:

You put $1,000 in the bank. The bank pays an annual rate of 8% interest, compounded quarterly. You leave your money there for 2 quarters. How much money do you have in your account at the end of 2 quarters?

Solution:

What I'd like you to do is plug this information into the formula. Then, we'll check to see that you did it correctly. You should have gotten:

$$A = P(1 + r)^t$$
$$A = \$1{,}000(1 + .02)^2$$

Let's go over just that. The initial principal is $1,000. The annual rate of interest is 8%, which comes to 2% per quarter for 2 quarters, $t = 2$.

The rest is up to you. Go ahead and solve for A.

$$\begin{aligned} A &= \$1{,}000(1 + .02)^2 \\ &= \$1{,}000(1.02)^2 \\ &= \$1{,}000(1.0404) \\ &= \$1{,}040.40 \end{aligned}$$

We carried out two arithmetic operations here: multiplication with decimals and quick multiplication. Now, just suppose you happened to have forgotten how to do either of these two operations. Where would you be able to review them? You'll find instructions on multiplication with decimals in chapter 6, frame 1, and you'll find information on quick multiplication in chapter 5, frame 1.

Problem 2:

Five hundred dollars is lent at an annual rate of 12% interest for 1 quarter. The interest is compounded monthly. How much money does the lender end up with (principal plus interest)?

Solution:

$$
\begin{aligned}
A &= P(1+r)^t \\
&= \$500(1+.01)^3 \\
&= \$500(1.01)^3 \\
&= \$515.15
\end{aligned}
$$

$$
\begin{array}{r}
1.01 \\
\times\, 1.01 \\
\hline
101 \\
1\,010 \\
\hline
1.0201 \\
\times\, 1.01 \\
\hline
10201 \\
1\,02010 \\
\hline
1.030301 \\
\times\, \$500 \\
\hline
\$515.150500
\end{array}
$$

Here's another one.

Problem 3:

Ten thousand dollars is lent at an annual rate of 16% for 1 year. The interest is compounded quarterly. How much money does the lender end up with at the end of the year?

Solution:

$$
\begin{aligned}
A &= P(1+r)^t \\
&= \$10{,}000(1+.04)^4 \\
&= \$10{,}000(1.04)^4 \\
&= \$10{,}000 \times 1.16985856 \\
&= \$11{,}698.59
\end{aligned}
$$

$$
\begin{array}{r}
1.04 \\
\times\, 1.04 \\
\hline
416 \\
1\,040 \\
\hline
1.0816 \\
\times\, 1.0816 \\
\hline
64896 \\
10816 \\
86528 \\
1\,08160 \\
\hline
1.16985856
\end{array}
$$

We pulled a couple of fast ones here. First, we multiplied 1.0816 by 1.0816 to save a step. We could have multiplied 1.0816 by 1.04 and *that* product by 1.04. But, after all, 1.04×1.04 is $(1.04)^2$, so we took a shortcut. Then, we multiplied 1.16985856 by \$10,000 by just moving the decimal four places to the right.

Before you begin Self-Test 2, ask yourself one question: "Do I really understand what's coming down—or, what's going up, for that matter?" If the answer's no, then you know where to go. You go right back to frame 2.

SELF-TEST 2

1. One thousand dollars is put in a bank that pays interest at an annual rate of 8%, compounded quarterly. How much money would be in the account after 3 quarters?
2. Ten thousand dollars is lent out at 4% annual rate for 1 year. If interest is compounded quarterly, how much money would the borrower owe the lender after 1 year?
3. One hundred thousand dollars is deposited in the bank. If the bank pays an annual rate of interest of 12%, compounded monthly, how much money is in the account after 4 months?

ANSWERS

1.
$$A = P(1+r)^t$$
$$= \$1{,}000(1+.02)^3$$
$$= \$1{,}000(1.02)^3$$
$$= \$1{,}000 \times 1.061208$$
$$= \$1{,}061.21$$

$$\begin{array}{r} 1.02 \\ \times 1.02 \\ \hline 204 \\ 1\,020 \\ \hline 1.0404 \\ \times 1.02 \\ \hline 20808 \\ 1\,04040 \\ \hline 1.061208 \end{array}$$

2.
$$A = P(1+r)^t$$
$$= \$10{,}000(1+.01)^4$$
$$= \$10{,}000(1.01)^4$$
$$= \$10{,}000 \times 1.04060401$$
$$= \$10{,}406.04$$

$$\begin{array}{r} 1.01 \\ \times 1.01 \\ \hline 101 \\ 1\,010 \\ \hline 1.0201 \\ \times 1.0201 \\ \hline 10201 \\ 204020 \\ 1\,02010 \\ \hline 1.04060401 \end{array}$$

3.
$$A = P(1+r)^t$$
$$= \$100{,}000(1+.01)^4$$
$$= \$100{,}000(1.01)^4$$
$$= \$100{,}000 \times 1.04060401$$
$$= \$104{,}060.40$$

Note: We can use the calculation from the second problem for $(1.01)^4$.

3 Doubling Time: The Rule of 70

No, this is not a variation of fast running or double time. It means the time it takes for a number to double.

How long would it take a nation's population to double if it is increasing at the rate of 1% a year? One hundred years, right? Not necessarily.

When someone mentions a rate of increase, you must immediately hold up your right hand, palm extended outward, and shout as loud as you can: "Hold it right there!"

Why such a strident reaction? Because someone may be trying to put something over on you. This person will look at you questioningly. Then you ask: "Are you talking about a simple 1% annual rate of increase of this year's base population, or are you talking about a 1% rate compounded annually?"

This query will immediately throw your adversary off balance, so you have the opportunity to answer your own question: "If you're talking about 1% compounded annually, the population will double in just 70 years. But if you're talking 1% of the base population—which is analogous to a simple 1% annual rate of increase—then it does take 100 years for the population to double."

How come we're so sure it takes exactly 70 years for the population of a country to double if it increases at a 1% compound annual rate? It actually takes 69.7 years, but who's counting? Besides, this gives us a chance to invoke the rule of 70, which certainly sounds a lot more punchy than the rule of 69.7.

Problem 1:

If you put $1,000 in the bank at 2% interest, compounded annually, how long would it take your money to double?

Solution:

It would take 35 years. That's right. All we need to do to find the doubling time is to divide the annual compound rate of increase into 70. So we divide 2 into 70, and we get 35.

Problem 2:

If your body weight were increasing at an annual compound rate of 7% per year, how long would it take for your weight to double?

Solution:

$70 \div 7 = 10$

So, it would take 10 years.

There is another way of determining the doubling time: $P(1+r)^t$. To do the first problem—how long would it take $1,000 to double if it were to increase at an annual compound rate of 2%—we'd need to fill in the formula. Go ahead and do that.

Did you get $1,000(1 + .02)^{35}$? How long would it take you to carry out $(1.02)^{35}$?

So the rule of 70 can come in quite handy. And it can, used judiciously, make you look like a whiz with numbers.

SELF-TEST 3

How much is the doubling time for each of the following problems?

1. A compound annual rate of 10%
2. A compound annual rate of 5%
3. A compound annual rate of 1%
4. A compound annual rate of 14%

ANSWERS

1. 7 years
2. 14 years
3. 70 years
4. 5 years

4 Discounting

Very often banks discount loans by deducting the interest owed from the face, or maturity, value of the loan.

Problem 1:

You take out a $1,000 loan at 7% interest. The bank gives you $930, which is $1,000 less the $70 interest (7% of $1,000 is $70). It does not take a mental giant to figure out what the bank is up to. It's up to charging you more than 7% interest. How much more? Figure it out.

Solution:

$$\text{interest rate} = \frac{\text{amount of interest}}{\text{loan amount} - \text{amount of interest}}$$

$$\frac{\$70}{\$930} = \frac{7}{93}$$

$$= 7.5\%$$

$$\begin{array}{r} .0752 \\ 93\overline{)7.0000} \\ -6.51^{xx} \\ \hline 490 \\ -465 \\ \hline 250 \\ -186 \\ \hline \end{array}$$

Now this may not seem like such a big discrepancy, but remember that banks compete by quarter-percent differences in interest rates. Furthermore, a difference of half a percent can really mean a lot more interest on a large loan. Take a 1-year, $100 million loan. Why, it's the difference between $____ and $____ in interest payments. You fill in the blanks.

Did you get $7.5 million and $7.6 million, or a difference of $100,000? Of course, if that amount of money is trivial to you, then perhaps you wouldn't mind sending me $100,000.

By the way, if finding the difference between $7.5 million and $7.6 million threw you, help is on the way. The next chapter is devoted to familiarizing you with big numbers—millions, billions, and trillions.

Problem 2:

How much is the actual interest rate on a $4,000 loan that a bank discounts at 12%?

Solution:

$$\frac{\$480}{\$3{,}520} = \frac{48}{352} = \frac{12}{88} = \frac{3}{22} = 13.6\%$$

```
      .136
 22 )3.000
   -2 2xx
      80
     -66
      140
     -132
        8
```

SELF-TEST 4

What is the actual interest rate on each of the following bank discounts? Each loan is for 1 year.

1. A $5,000 loan discounted at 8%
2. A $2,000 loan discounted at 10%
3. A $10,000 loan discounted at 7%

ANSWERS

1. $\frac{\$400}{\$4{,}600} = \frac{4}{46} = \frac{2}{23} = 8.7\%$

```
      .087
 23 )2.000
    -1 84x
      160
```

2. $\frac{\$200}{\$1{,}800} = \frac{2}{18} = \frac{1}{9} = 11.1\%$

```
     .1 1 1 1
 9 )1.0¹0¹0¹0
```

3. $\frac{\$700}{\$9{,}300} = \frac{7}{93} = 7.5\%$

```
      .075
 93 )7.000
    -6 51x
      490
     -465
       25
```

5 The True Rate of Interest

We saw in the last section that when banks discount loans, the borrowers end up paying somewhat more interest than the interest rate represented by the banks. And yet, if the borrowers repay their loans in monthly installments, they have the use of smaller and smaller pieces of what they had originally borrowed.

Suppose you were to borrow \$1,200 at 10% interest on January 1. On a discount loan, the bank would actually give you \$1,080 and tell you to pay back the money in 12 monthly installments of \$100 on the first day of each month, beginning in February. On February 1, you'll send the bank a check for \$100, which leaves you with just \$980 (\$1,080 − \$100). On March 1, you're down to \$880. You can see that that so-called \$1,200 loan is shrinking pretty fast.

So I'll tell you what I'm gonna do. I'm gonna show you what you're really paying for that bank loan. What you're paying is called the true rate of interest.

$$i = \frac{2MC}{P(N+1)}$$

The letter *i* represents the true rate of interest. *M* is the length of payment period divided into 1 year (e.g., when payments are quarterly, $M = 4$; when they are monthly, $M = 12$). *C* represents the total dollar amount of charges for the loan (interest, carrying, and finance charges). *P* is the principal of the loan (the amount actually received). *N* is the number of installment repayments made.

Problem 1:

We'll go over the \$1,200 bank loan discounted at 10% that is paid off in 12 monthly installments. Write down the formula, substitute for *M, C, P,* and *N* in the formula, and then solve for *i*.

Solution:

$$i = \frac{2MC}{P(N+1)} = \frac{2 \times 12 \times \$120}{\$1{,}080(12+1)} = \frac{24 \times \$120}{\$1{,}080(13)}$$

$$= \frac{24 \times \$12\cancel{0}}{\$1\cancel{0}80\ (13)}_{9}$$

$$= \frac{24}{117}$$

$$= 20.51\%$$

$$\begin{array}{r} .2051 \\ 117\overline{)24.0000} \\ -23\,4^{xxx} \\ \hline 600 \\ -585 \\ \hline 150 \end{array}$$

Isn't it amazing how what appears to be a straightforward 10% bank loan is, in truth, really a 20.51% loan? Is it any wonder that banks and other lending institutions have always opposed truth-in-lending laws? Loan sharks may charge higher interest rates, but at least they let you know what they're charging, not to mention what will happen to you if you don't keep up your payments.

Problem 2:

What is the true rate of interest on a $5,000 bank loan, discounted at 8%, with quarterly payments for 1 year?

Solution:

$$i = \frac{2MC}{P(N+1)} = \frac{2 \times 4 \times \$400}{\$4,600(4+1)} = \frac{8 \times 4}{46(5)} = \frac{16}{23 \times 5} = \frac{16}{115} = 13.9\%$$

$$\begin{array}{r} .139 \\ 115\,\overline{)16.0000} \\ -11\,5^{xxx} \\ \hline 4\,50 \\ -3\,45 \\ \hline 1\,050 \\ -1\,035 \\ \hline 150 \end{array}$$

Problem 3:

What is the true rate of interest on a $10,000 bank loan, discounted at 12%, with monthly payments for 1 year?

Solution:

$$i = \frac{2MC}{P(N+1)} = \frac{2 \times 12 \times \$1,200}{\$8,800(12+1)} = \frac{24 \times 12}{88 \times 13} = \frac{24 \times 3}{22 \times 13} = \frac{72}{286} = \frac{36}{143} = 25.2\%$$

$$\begin{array}{r} 22 \\ \times 13 \\ \hline 66 \\ 22 \\ \hline 286 \end{array}$$

$$\begin{array}{r} .2517 \\ 143\,\overline{)36.0000} \\ -28\,6^{xx} \\ \hline 7\,40 \\ -7\,15 \\ \hline 250 \\ -143 \\ \hline 1070 \\ -1001 \\ \hline \end{array}$$

It's very easy to make arithmetic mistakes here, because the formula for the true rate of interest is pretty complex. If you feel you understand how to set up and solve these problems, go on to Self-Test 5. If you don't, then go back to frame 5.

SELF-TEST 5

Find the true rate of interest for each of the following loans.

1. A $20,000 bank loan, discounted at 10%, with monthly payments for 1 year.
2. A $5,000 bank loan, discounted at 9%, with quarterly payments for 1 year.
3. A $10,000 bank loan, discounted at 14%, with monthly payments for 1 year.

ANSWERS

1. $$i = \frac{2MC}{P(N+1)} = \frac{2 \times 12 \times \$2{,}000}{\$18{,}000(12+1)} = \frac{24}{9 \times 13} = \frac{24}{117} = 20.5\%$$

$$\begin{array}{r} .205 \\ 117\overline{)24.000} \\ -23\,4^{xx} \\ \hline 600 \\ -585 \\ \hline \end{array}$$

2. $$i = \frac{2MC}{P(N+1)} = \frac{2 \times 4 \times \$450}{\$4{,}550(4+1)} = \frac{8 \times \$45}{\$455 \times 5} = \frac{8 \times 9}{455} = \frac{72}{455} = 15.8\%$$

$$\begin{array}{r} .1582 \\ 455\overline{)72.0000} \\ -45\,5^{xxx} \\ \hline 26\,50 \\ -22\,75 \\ \hline 3\,750 \\ -3\,640 \\ \hline 1100 \\ -910 \\ \hline 190 \end{array}$$

3. $$i = \frac{2MC}{P(N+1)} = \frac{2 \times 12 \times \$1{,}400}{\$8{,}600(12+1)} = \frac{24 \times 14}{86 \times 13} = \frac{12 \times 14}{43 \times 13} = \frac{168}{559} = 30.1\%$$

$$\begin{array}{r} 14 \\ \times 12 \\ \hline 28 \\ 14 \\ \hline 168 \end{array} \qquad \begin{array}{r} 43 \\ \times 13 \\ \hline 129 \\ 43 \\ \hline 559 \end{array} \qquad \begin{array}{r} .3005 \\ 559\overline{)168.0000} \\ -167\,7^{xxx} \\ \hline 3000 \\ -2795 \\ \hline \end{array}$$

20 Big Numbers

The main purpose of this book is to put you at ease with numbers. So far, we've concentrated on relatively small numbers—fractions, decimals, hundreds of dollars, thousands of dollars—although in the last chapter, we soared to $7.6 million.

Well, how do you feel about *very* large numbers? Billions? Trillions? Quadrillions? Quintillions? Can you divide four hundred thirty-two quadrillion by one hundred and eight trillion? Did you get four thousand? If you did, then you definitely know your big numbers and can move all the way around the board, pass GO, and collect $200. In fact, you should go all the way to chapter 21.

1 Thousands and Millions

If you won a million dollars in the lottery, you would definitely know how to count it. Everyone can count well past one million. And if inflation picks up again, we may well have to.

Problem 1:
Write out the number one million—with all those zeros.

Solution:
1,000,000

Now we'll do a much easier one.

Problem 2:

Write out the number one hundred fifty-three thousand.

Solution:

153,000

Now let's compare thousands and millions. One hundred fifty-three thousand is the number one hundred fifty-three followed by a comma and three zeros. So the numbers to the left of that comma and three zeros signify thousands.

Next come millions. One million is designated by a one, a comma, three zeros, another comma, and three more zeros. So one million is a one and two sets of three zeros.

Problem 3:

Please write the number eight hundred and four million.

Solution:

804,000,000

Problem 4:

Please write the number two hundred sixteen million, four hundred seventy-six thousand.

Solution:

216,476,000

Now we'll reverse the process.

Problem 5:

Please express this number in words: 547,302.

Solution:

Five hundred forty-seven thousand, three hundred and two

Problem 6:

Express 81,963,000 in words.

Solution:

Eighty-one million, nine hundred sixty-three thousand

Problem 7:

Express this number in words: 905,100,515.

Solution:

Nine hundred and five million, one hundred thousand, five hundred and fifteen

SELF-TEST 1

Please translate these words into numbers:

1. Four hundred seventy-five thousand, five hundred
2. Two million
3. Seven hundred seventy-four million, two hundred fifty thousand
4. Ninety-three thousand and two
5. Three hundred forty-six million, five hundred sixty-one thousand, two hundred eighty-eight

Please express these numbers in words:

6. 75,000
7. 400,160,000
8. 145,005
9. 785,631,072
10. 100,175,200

ANSWERS

1. 475,500
2. 2,000,000
3. 774,250,000
4. 93,002
5. 346,561,288
6. seventy-five thousand
7. four hundred million, one hundred sixty thousand
8. one hundred forty-five thousand and five
9. seven hundred eighty-five million, six hundred thirty-one thousand, and seventy-two
10. one hundred million, one hundred seventy-five thousand, two hundred

2 Billions and Trillions

You may win a million dollars in the lottery, but I'd be willing to give you a million-to-one odds that you'll never win a billion. In fact, a lot of people can't even write out the number one billion with all those zeros. Can you?

One billion is the number one, followed by three sets of zeros, three zeros to a set: 1,000,000,000

Is there any use for this number? Well, for starters, the government spends nearly $300 billion a year on defense, and over $400 billion on Social Security. Political candidates toss billion-dollar figures around like Super Bowl quarterbacks toss footballs.

Problem 1:

Write out the number seven hundred billion, three hundred and fifty-five million.

Solution:

700,355,000,000

Problem 2:

And now, just to make sure we're both on the same wavelength, write out the number four hundred and three billion, two hundred nineteen million, seven hundred forty-two thousand, and one.

Solution:

403,219,742,001

Problem 3:

How would you write one trillion, with all its zeros?

Solution:

1,000,000,000,000

Did you put down a one and four sets of zeros, three to a set?

Problem 4:

How about writing out our 1997 Gross Domestic Product, or GDP, which came to $8.1 trillion?

Solution:

Our 1997 GDP was $8,100,000,000,000. In other words, eight trillion, one hundred billion dollars.

Problem 5:

Now write out this number: five hundred forty-three trillion, two hundred ninety-six billion, five hundred million.

Solution:

543,296,500,000,000

Let's turn things around and put numbers into words.

Problem 6:

Express 7,500,000,000 in words.

Solution:

Seven billion, five hundred million

Problem 7:

Now express 511,388,950,000,000 in words.

Solution:

Five hundred eleven trillion, three hundred eighty-eight billion, nine hundred and fifty million.

How's it going? If none of this throws you, do Self-Test 2. But if you're still not comfortable with millions, billions, and trillions, go back to frame 1. It's often better the second time around.

SELF-TEST 2

Please translate these words into numbers:

1. fifty billion
2. four hundred trillion, two hundred and thirty billion
3. seventy-two trillion, nine hundred fifty-four billion, three hundred and eight million
4. ten trillion
5. five hundred and two trillion, four hundred sixty-four billion, nine hundred and seventeen million

Please express these numbers in words:

6. 175,000,000,000
7. 800,000,000,000
8. 675,505,768,400,000
9. 3,000,000,000,000
10. 234,459,875,361,827

ANSWERS

1. 50,000,000,000
2. 400,230,000,000,000
3. 72,954,308,000,000
4. 10,000,000,000,000
5. 502,464,917,000,000
6. one hundred and seventy-five billion
7. eight hundred billion
8. six hundred seventy-five trillion, five hundred and five billion, seven hundred sixty-eight million, four hundred thousand
9. three trillion
10. two hundred thirty-four trillion, four hundred fifty-nine billion, eight hundred seventy-five million, three hundred sixty-one thousand, eight hundred and twenty-seven

3 Quadrillions, Quintillions, and Even Bigger Numbers

Here come some really big numbers. These numbers are so big that except during times of runaway inflation—like the one in Hungary after World War II, when you needed 828 octillion pengos to buy what 1 pengo bought before the war—you'll probably never encounter them.

Then why bother with these numbers? They are useful in helping to illustrate the basis for our numbering system.

Let's use the number 17,550,819,346,122 as an illustration. It is read seventeen trillion, five hundred fifty billion, eight hundred nineteen million, three hundred forty-six thousand, one hundred and twenty-two. There's a message hidden somewhere in these numbers.

One hundred twenty-two is a hundreds number. Hundreds numbers are to the right of the first comma. Thousands numbers—in this case, 346—are to the right of the second comma.

Once we have those first two commas out of the way, we can begin counting sets of digits. The next set of digits, 819, represents millions. The prefix "mil" means one. Next comes 550, or five hundred fifty billion. The prefix "bi," of course, means two. You know—bicycle, bicentennial, biweekly, and so on.

And now, everything else should fall into place—17 represents seventeen trillion. And "tri" means three. Next comes quadrillions (four), quintillions (five), sextillions (six), septillions (seven), and octillions (eight).

Don't worry about these really big numbers. You'll never use them. So, as a special treat, there's no self-test at the end of this section. However, please examine the box on Millions, Billions, Trillions, and Beyond for more on these large numbers.

Millions, Billions, Trillions, and Beyond	
one million	1,000,000
one billion	1,000,000,000
one trillion	1,000,000,000,000
one quadrillion	1,000,000,000,000,000
one quintillion	1,000,000,000,000,000,000
one million is $1{,}000^2$	
one billion is $1{,}000^3$	
one trillion is $1{,}000^4$	
one quadrillion is $1{,}000^5$	
one quintillion is $1{,}000^6$	

4 Multiplying Big Numbers

How much is $1{,}000 \times 1{,}000$? When we multiply any whole number (i.e., not a fraction or decimal) by 1,000, we just add three zeros, so $1{,}000 \times 1{,}000 = 1{,}000{,}000$.

How much is 1,000 × \$1,000,000? It comes to \$1,000,000,000, or one billion dollars.

And finally, what's $1{,}000 \times 1{,}000{,}000{,}000$? It's 1,000,000,000,000, or one trillion. So one thousand thousand is one million. One thousand million is one billion. And one thousand billion is one trillion. Now we'll deal with numbers that don't all end in zero.

Problem 1:

How much is $10{,}000 \times 1{,}475$?

Solution:

14,750,000

Problem 2:

How much is $100{,}000 \times 1{,}302{,}116$?

Solution:

130,211,600,000

What it all comes down to when we multiply very large numbers is adding zeros. There's a temptation to memorize rules: When we multiply by 10, we add one zero; when we multiply by 100, we add two zeros;

when we multiply by 1,000, we add three zeros, and so forth. But why bother to memorize all these rules, when all we need to do is look? That's right—if we multiply by 1,000, add three zeros. Why? Because they're there. Just look at the number you're multiplying by and add its zeros to the number being multiplied. Of course, you do have to carry out the rest of the multiplication.

Problem 3:

How much is 16,000 × 4,000?

Solution:

16 × 4 = 64 + 6 zeros = 64,000,000

Problem 4:

How much is 134,100 × 200,000?

Solution:

1,341 × 2 = 2,682 + 7 zeros = 26,820,000,000

5 Dividing Big Numbers

Division is the reverse of multiplication. Dividing by 1,000 means subtracting three zeros. Twenty million divided by one thousand equals twenty thousand (20,000,000 becomes 20,000).

Problem 1:

Divide 275 billion by one thousand.

Solution:

275 million (275,000,000,000 becomes 275,000,000)

Similarly, dividing by 10,000 means subtracting four zeros, and dividing by 100,000 means subtracting five zeros.

Problem 2:

How much is two million divided by ten thousand?

Solution:

2,000,000 ÷ 10,000 = 200

Problem 3:

How much is forty billion divided by one hundred thousand?

Solution:

$40{,}000{,}000{,}000 \div 100{,}000 = 400{,}000$

Let's put all these great skills to some practical use. Suppose that we divided up all our production, valued at current prices, and gave every man, woman, and child in this country the same share. How much would that come to? In other words, how much is the per-capita GDP of the United States? This is not the type of question you hear every day, but it's a great way to illustrate division with big numbers.

Problem 4:

Our per-capita GDP, or income per person, is found by dividing our GDP by our population. Suppose that our GDP were \$5 trillion and our population, 250 million. Calculate our per-capita GDP.

Solution:

$$\text{per-capita GDP} = \frac{\text{GDP}}{\text{population}}$$

$$= \frac{\$5{,}000{,}000{,}000{,}000}{250{,}000{,}000}$$

Eliminate seven zeros from the top and seven from the bottom:

$$\frac{\$500{,}000}{25} = \frac{\$100{,}000}{5} = \$20{,}000$$

Problem 5:

Now calculate the per-capita GDP of China, assuming its GDP is \$1 trillion and its population is one billion.

Solution:

$$\text{per-capita GDP} = \frac{\text{GDP}}{\text{population}}$$

$$= \frac{\$1{,}000{,}000{,}000{,}000}{1{,}000{,}000{,}000} = \$1{,}000$$

There's a shortcut. You may recall that one trillion has three more zeros than one billion, or that one trillion is one thousand billion. So, either way, the answer is $1,000.

SELF-TEST 3

1. How much is one thousand thousand?
2. How much is a thousand billion?
3. How much is $1{,}000^4$?
4. How much is $10{,}000 \times 7{,}564$?
5. How much is $100{,}000 \times 56{,}412$?
6. How much is $12{,}000 \times 6{,}000$?
7. How much is $2{,}340 \times 200{,}000$?
8. Divide 340 million by 1,000.
9. Divide 15 billion by 10,000.
10. How much is 6.5 trillion divided by one hundred thousand?
11. If the European Economic Community has a total GDP of $10.5 trillion and its combined population is 350 million, how much would its per-capita GDP be?
12. What is the per-capita GDP of a nation with a GDP of $500 billion and a population of 200 million?

ANSWERS

1. 1,000,000
2. 1,000,000,000,000
3. 1,000,000,000,000
4. 75,640,000
5. 5,641,200,000
6. 72,000,000
7. 468,000,000
8. 340,000
9. 1,500,000
10. 65,000,000
11. $$\text{per-capita GDP} = \frac{\text{GDP}}{\text{population}} = \frac{\$10{,}500{,}\cancel{000}{,}\cancel{000}{,}\cancel{000}}{350{,}\cancel{000}{,}\cancel{000}} = \$30{,}000$$
12. $$\frac{\$500{,}\cancel{000}{,}\cancel{000}{,}\cancel{000}}{200{,}\cancel{000}{,}\cancel{000}} = \$2{,}500$$

21 A Taste of Statistics

Statistics can be lied with, doctored, and manipulated. But I promise to restrain myself, so you can relax.

We'll be covering just three topics—the mean, the median, and the mode. What's convenient about these topics is that they provide some practical uses for the math we've already covered. And best of all, you won't even need a calculator.

1 The Mean, the Median, and the Mode

The mean, the median, and the mode are three very simple statistical measures. The mean is the average of all the numbers used. The median is the middle number, and the mode is the number most often mentioned.

Here is a set of numbers:

2,5,7,8,8

The mean for these numbers is found by adding them all and then dividing by the number of terms: $30 \div 5 = 6$. So the mean is 6.

How much is the median? That one's easy. The median is 7, which is the middle term. Of course, the median is the middle term only when the numbers are in order. When they aren't, you must first put them in order of magnitude before you can determine the median.

And how about the mode? The mode is 8, because 8 is mentioned twice, while each of the other numbers is mentioned only once.

Are you ready for another example?

Problem 1:

Find the mean, median, and mode for this array:
6,9,5,8,9,5,9

Solution:

The mean is 51 ÷ 7 = 7.29, or 7.3. The median is found by putting the array in order: 5, 5, 6, 8, 9, 9, 9. The median, or middle number, is 8. The mode is 9, which is mentioned more frequently than any of the other numbers.

And now for a trick. What if there is more than one mode? What if there are *two* numbers mentioned with greater frequency than any other numbers? No problem. We'll call it a bimodal distribution and get on with our lives.

Problem 2:

In the following array, what are the modes?
8,4,1,6,3,8,2,1,7,3,8,2,1

Solution:

8 and 1

And now for something not all that different.

Problem 3:

What are the modes for the following array?
6,2,19,4,3,17,6,8,19,3,15

Solution:

They are 3, 6, and 19, which is a trimodal distribution. Don't bother trying to remember such fancy terms as bimodal and trimodal—unless you're into impressing people with your sophisticated vocabulary.

And now back to the median.

Problem 4:

In this array, what is the median?
8,10,11,12,15,16

Solution:

Since there is no middle number, we should take the average of two middle numbers, 11 and 12. And what is the average of 11 and 12? We add 11 and 12 to get 23 and divide that by 2, which gives us 11.5. So the median is 11.5.

SELF-TEST 1

1. Find the mean, median, and mode of this array: 31,16,22,19,17,22,15
2. Find the mean, median, and mode of this array: 27,14,2,9,4,13,14,7
3. Find the mean, median, and modes of this array: 4,5,9,4,8,1,10,5,1,3

ANSWERS

1. Mean = 142 ÷ 7 = 20.3; median = 19; mode = 22
2. Mean = 90 ÷ 8 = 11.3; median = 22/2 = 11; mode = 14
3. Mean = 50 ÷ 10 = 5; median = 9/2 = 4.5; modes = 1, 4, 5

22 Personal Finances

As individual consumers, we are called upon to use arithmetic and simple algebra almost every day. Sometimes we do mathematical calculations without even being conscious that we are doing them. And other times, we are all too aware that it is mathematics that we are trying to do.

In this chapter, we have taken up six types of numerical problems that most of us encounter in our day-to-day lives. They all have one thing in common—percentages. As you'll soon see, percentages are central to our daily existence. And even more to the point, everyone seems to be after a percentage of our money.

1 Mark-Down Problems

Everybody looks for bargains. Advertisements such as "Everything must go," "Prices slashed by 50% and more," and "Lost our lease" attract bargain hunters like honey attracts bees. Let's see how much you're actually saving.

Problem 1:

A dress is marked down from \$99 to \$59. By what percent has the price been cut?

Solution:

$$\text{percentage change} = \frac{\text{change}}{\text{original price}} = \frac{40}{99}$$

$$99\overline{)40.000} = .404 = 40.4\%$$

```
     .404 = 40.4%
99 )40.000
  −39 6xx
      400
     −396
```

Problem 2:

You will receive a $500 rebate on a $9,000 car. What percentage of the original price do you get back?

Solution:

$$\text{percentage change} = \frac{\text{rebate}}{\text{original price}} = \frac{\$500}{\$9{,}000}$$

```
9000 )500 = 90 )5
          .0555 = 5.6%
   = 18 )1.0000
         −90xx
          100
          −90
          100
          −90
```

Problem 3:

A pair of shoes originally priced at $39.95 was marked down by 20%. What is the new price?

Solution:

```
  $39.95            $39.95
   ×.20             −7.99
$7.9900 = $7.99     $31.96
```

How are you doing? If you have gotten at least two out of these three right, then go to the next problems. If not, then you should review frame 2 of chapter 10 and return to the beginning of this chapter.

And now for something a little different. And a little harder.

Problem 4:

A jacket is marked down by 25% to $50. What was its original price?

Solution:

Let x = the original price

$$x - .25x = \$50$$
$$.75x = \$50$$
$$x = \frac{\$50}{.75}$$

$$.75\overline{)\$50} = 75\overline{)\$5{,}000}$$
$$\begin{array}{l} \quad\; 6\;6.\;6\;6\;6 \\ = 3\overline{)20{}^{2}0.{}^{2}0.{}^{2}0{}^{2}0} \end{array}$$
$$= \$66.67$$

Problem 5:

A bicycle was marked down by 40% to \$120. What was its original price?

Solution:

Let x = the original price

$$x - .4x = \$120$$
$$.6x = \$120$$
$$x = \frac{\$120}{.6}$$

$$.6\overline{)\$120} = 6\overline{)\$1{,}200}$$
$$= 1\overline{)\$200}$$
$$= \$200$$

If you got both problems right, then go to Self-Test 1. And if you got them wrong, then you need to review some of your algebra. Remember x, and all the things you may let it represent? Return to frame 1 of chapter 18 and reread the entire chapter. Then go back to Problem 4 in this section.

SELF-TEST 1

1. A sofa is marked down from \$599 to \$299. By what percentage has the price been cut?
2. An auto dealer is offering a \$1,000 rebate on a \$14,000 car. What percentage of the original price do you get back?
3. A dress that was originally \$150 was marked down by 40%. What is the new price?
4. A living room set is marked down by 35% to \$750. What was its original price?
5. A bedroom set was marked down by 60% to \$975. What was its original price?

ANSWERS

1. $\text{percentage change} = \dfrac{\text{change}}{\text{original price}} = \dfrac{\$300}{599} = 50.1\%$

$$599\overline{)300.0000} = .5008$$
$$\begin{array}{r} 300.0000 \\ -299\,5^{xxx} \\ \hline 5000 \\ -4792 \\ \hline 208 \end{array}$$

2. $\text{percentage change} = \dfrac{\$1{,}000}{14{,}000} = \dfrac{1}{14} = 7.1\%$

$$14\overline{)1.000} = .071$$
$$\begin{array}{r} 1.000 \\ -98^{x} \\ \hline 20 \\ -14 \\ \hline 6 \end{array}$$

3. $$\begin{array}{r} \$150 \\ \times .4 \\ \hline \$60 \end{array} \qquad \begin{array}{r} \$150 \\ -60 \\ \hline \$90 \end{array}$$

4. Let x = original price

$$x - .35x = \$750$$
$$.65x = \$750$$
$$x = \frac{\$750}{.65}$$
$$= \$1{,}153.85$$

$$.65\overline{)\$750} = 65\overline{)\$75{,}000} = 13\overline{)\$15000.000} = \$1153.846$$
$$\begin{array}{r} 15000.000 \\ -13^{xxx\ xxx} \\ \hline 20 \\ -13 \\ \hline 70 \\ -65 \\ \hline 50 \\ -39 \\ \hline 11\,0 \\ -10\,4 \\ \hline 60 \\ -52 \\ \hline 80 \\ -78 \end{array}$$

5. Let x = original price

$$x - .6x = \$975$$
$$.4x = \$975$$
$$4x = \$9{,}750$$
$$x = \frac{\$9{,}750}{4}$$
$$= \$2{,}437.50$$

$$4\overline{)\$9^{1}7^{1}5 3 0.^{2}0} = \$2\,4\,3\,7.\,5$$

Which Is the Better Deal?

Who is your long distance carrier? Maybe I can get you to switch.

Friendly Fhones offers you a flat rate of 10 cents a minute, any time, day or night. Chat Line claims to have an even better deal—12 cents a minute, any time, day or night. But your first 30 minutes each month are free. Who is offering the better deal? See if you can work it out.

Solution:

Using Friendly Fhones, your first 30 minutes cost you \$3.00 ($30 \times \$.10$). But your first 30 minutes are free using Chat Line. So if you talk for only 30 minutes a month, Chat Line is definitely the way to go.

But after the first 30 minutes, Chat Line costs you 12 cents a minute, while Friendly Fhones is just 10 cents a minute. So for every additional minute you use Friendly Fhones, you're saving 2 cents. How long beyond the initial 30 minutes would you need to talk to make up the entire $3?

The answer is 150 minutes. Okay, then, which long distance carrier gives you the better deal?

If you talk for more than three hours (30 minutes plus 150 minutes), then Friendly Fhones is the better deal. But if you talk less than three hours a month, you should definitely go with Chat Line.

2 Sales Tax Problems

When you pay sales tax, you never have to worry about calculating it because that's the job of the seller. But you're the one who pays the tax. In fact, most people don't even pay attention to how much they're paying for the good or service and how much they're being charged in sales tax. In New York City, for example, there's a sales tax of 8 1/4% on most items—clothing, restaurant meals, books, furniture, movie admissions. In New Jersey, just across the river, they charge only 5%, and clothing is tax exempt. So maybe it pays to drive over to New Jersey to shop at Daffy Dan's or at the Secaucus retail outlets. Not only do you save on your sales tax, but you can pick up some real bargains. Let's figure out how much you would save.

Problem 1:

If a dress in Bloomingdale's were priced as $129 and Daffy Dan's had the same dress for $89, how much would you save? Did you say $40? Guess again. You would save even more, because you need to figure out the sales tax that you would have paid in New York (8 1/4%). In New Jersey, there's no sales tax on clothing.

Solution:

```
    $129
  ×.0825
     645
    258
  10 32
$10.6425
```

You would be saving \$10.64 in tax, in addition to the \$40 price difference. So you would save a total of \$50.64. On the other hand, you'd have to pay for gas and tolls.

Problem 2:

If there is a sales tax of 6%, how much tax would you pay on a used car that was priced at \$2,500?

Solution:

$$\begin{array}{r} \$2{,}500 \\ \underline{\times .06} \\ \$150.00 \end{array}$$

Now we'll get a bit more fancy.

Problem 3:

How much would the original price of an item be if you paid a total (including taxes) of \$102.90 and the sales tax rate was 5%?

Solution:

First, you need to find the price of the item that you purchased before the tax was added.

Let $x =$ the original price

$$x + .05x = \$102.90$$
$$1.05x = \$102.90$$
$$x = \frac{\$102.90}{1.05}$$
$$= \$98$$

$$1.05\overline{)102.90} = 105\overline{)10290}$$
$$\begin{array}{r} 98 \\ = 21\overline{)2058} \\ \underline{-189^{x}} \\ 168 \\ \underline{-168} \end{array}$$

Problem 4:

How much would the original price be if you paid a total (including taxes) of \$520 for a sofa and the sales tax rate were 4%?

Solution:

Let $x =$ original price of sofa

$$x + .04x = \$520$$
$$1.04x = \$520$$

$$x = \frac{\$520}{1.04}$$

$$= \$500$$

$104\overline{)52000} = 26\overline{)13000}$ with quotient 500; -130

If you got these last two right, go to Self-Test 2. If you got them wrong, please turn to the beginning of chapter 18, which is a review of algebra, and read the entire chapter. The whole trick with these problems is to let x represent the unknown. Then everything else will fall into place. When you've finished chapter 18, return to frame 2 of this chapter.

SELF-TEST 2

1. If the sales tax on a \$450 purchase were 4 1/2%, how much tax would you pay?
2. If the sales tax rate in North Dakota were 7% and the sales tax in South Dakota were 3%, how much money would a person save by buying a \$9,000 car in South Dakota?
3. How much would the original price be if you paid a total (including taxes) of \$350 for a kitchen set and the sales tax rate were 5%?
4. How much would the original price be if you paid a total (including taxes) of \$94.50 for a dress and the sales tax were 5%?

ANSWERS

1.

$$\begin{array}{r} \$450 \\ \times .045 \\ \hline 2\,250 \\ 18\,00 \\ \hline \$20.250 \end{array}$$

2.

$$\begin{array}{r} \$9000 \\ \times .04 \\ \hline \$360 \end{array}$$

3. Let x = original price

$$x + .05x = \$350$$
$$1.05x = \$350$$
$$x = \frac{\$350}{1.05}$$
$$= \$333.33$$

$1.05\overline{)350} = 105\overline{)35000}$

$= 21\overline{)7000}$

$= 3\overline{)1000}$ with quotient \$333.33

4. Let x = original price

$$x + .05x = \$94.50$$
$$1.05x = \$94.50$$
$$x = \frac{\$94.50}{1.05}$$
$$= \$90$$

$1.05\overline{)94.50} = 105\overline{)9450}$

$= 21\overline{)1890}$ with quotient 90; -189

3 Credit Card Problems

The following section is written as a public service, especially for credit card holders who are being ripped off by high interest rates. Do you ever read the fine print on the back of your bill? Do you know how much interest you are being charged on your unpaid balance? Well, you're about to find out.

Suppose you make purchases this month totaling $1,050 dollars and the bank asks you to make a minimum payment of $50. So what do you do? You pay $50. Then the bank charges you, say, 17.5% interest on the unpaid balance. This annual rate is translated into a monthly rate of 1.458%.

Problem 1:

How much interest do you pay?

Solution:

$$\begin{array}{r} .01458 \\ \times \$1000 \\ \hline \$14.58000 = \$14.58 \end{array}$$

Remember how to multiply by 1,000? Just add three zeros to the number you're multiplying, or, what amounts to the same thing, move that number's decimal point three places to the right. This gives us $14.58, so you just paid $14.58 for the privilege of carrying a $1,000 balance on your credit card for 1 month.

Problem 2:

You have an outstanding monthly balance of $2,075, and there's a minimum payment due of $75. If the annual interest rate on your balance is 19% and you pay just the minimum, how much interest do you owe after 1 month?

Solution:

$$\text{monthly interest rate} = \frac{\text{annual rate}}{12} = \frac{19\%}{12} = 1.583\%$$

$$\begin{array}{r} 1.583 \\ 12\overline{)19.000} \\ -12 \\ \hline 7\,0 \\ -6\,0 \\ \hline 1\,00 \\ -96 \\ \hline 40 \\ -36 \\ \hline \end{array}$$

$$\begin{array}{r} .01583 \\ \times \$2000 \\ \hline \$31.66000 = \$31.66 \end{array}$$

Problem 3:

If you owe $47 interest on a monthly credit card balance of $3,000, what is the annual and monthly interest rates that you must pay?

Solution:

$$\frac{\$47}{\$3{,}000} = 3{,}000\overline{)47.000}$$

$$3\overline{).04^{1}7^{2}0^{2}0^{2}0} \quad (\text{quotient } .01\,5\,6\,6\,6)$$

$$= 1.567\% \text{ monthly}$$

$$\begin{array}{r} 1.567 \\ \times 12 \\ \hline 3\,134 \\ 15\,67 \\ \hline 18.804 \end{array}$$

$18.804 = 18.8\%$ annually

Problem 4:

If you owe $64 interest on a monthly credit card balance of $4,000, what is the annual and monthly interest rates you must pay?

Solution:

$$\frac{\$64}{\$4{,}000} = \frac{\$16}{\$1{,}000}$$

$$= 1.6\% \text{ monthly}$$

$$\begin{array}{r} 1.6 \\ \times 1\,2 \\ \hline 3\,2 \\ 16 \\ \hline \end{array}$$

19.2% annually

To put all of this into perspective, if you happened to have a bank balance of $1,000 and you had an average credit card balance of $1,000, you would end up paying 16% or 18% interest on your card debt while receiving just 5% interest for your bank deposit. And if you maintained an account at the same bank that issued your credit card, you would have the privilege of borrowing your own money and paying somewhere between 11% and 13% interest for this privilege. That's real smart! (See the box on the Passbook Loan Scam for a further discussion of this topic.)

Did you know that the banks whose credit cards you hold don't want you to ever pay off your balances? After all, they're charging you 16% or 18% interest on money that they borrow at about 5%. And often they have even borrowed that money from you. Borrow cheap and lend dear—that's what banking is all about.

The Passbook Loan Scam

Bankers like nothing more than lending you back your own money at a higher interest rate than they're paying you for it. Or to modify a popular Frank Perdue chicken ad, it takes a dumb borrower to take out a dumb loan.

Here's a typical example of a passbook loan. The bank is paying you 5% interest on your passbook savings. Normally a borrower would be charged 12% for a personal loan. But because you are required to leave the amount of the loan in your savings account, the bank will allow you to borrow this amount at a much lower rate, say, 8%.

Have you followed so far? The bank usually charges borrowers 12% interest on personal loans, but because you're such a respected customer—and you're going to leave the amount that you're borrowing right there in your savings account—the bank is willing to lend you this money at just 8% interest. And, the bank may add, your money is still earning 5% interest since you were able to leave it in your account while the bank lent you the money.

Consider the alternative. You withdrew the money instead of borrowing it from the bank. Say it happened to be $10,000. Eventually you would put it back into your account, just as you would have when you repaid the loan. So you could have had your own money to use. But by taking out the passbook loan, you, in effect, borrowed your own $10,000 and paid 3% interest (the 8% you paid less the 5% interest you received). If you borrowed the money for 1 year, this transaction would cost you 3% of $10,000, or $300. So, you have managed to borrow your own money and to have paid the bank $300. It takes a dumb borrower to take out a dumb loan.

SELF-TEST 3

1. How much interest would you pay on a credit card balance of $3,000 in 1 month if the annual rate of interest were 18%?
2. How much interest would you pay on a credit card balance of $2,500 if the annual rate of interest were 15%?

3. If you owe $50 interest on a monthly credit card balance of $3,300, what are the annual and monthly interest rates that you must pay?

4. If you owe $70 interest on a monthly credit card balance of $4,900, what are the annual and monthly interest rates that you must pay?

5. How much would it cost you in interest if you took out a 1-year passbook loan of $5,000 at 7% interest if the bank pays 5% interest on passbook savings? How much more does this loan cost you than if you had withdrawn $5,000 from the bank?

ANSWERS

1. $\frac{18\%}{12} = \frac{3}{2} = 1.5\%$

$$\begin{array}{r} \$3000 \\ \times .015 \\ \hline 15\,000 \\ 30\,00 \\ \hline \$45.000 = \$45.00 \end{array}$$

2. $\frac{15\%}{12} = 1\frac{3}{12} = 1\frac{1}{4}$

$$\begin{array}{r} 2500 \\ \times .0125 = 1.25\% \\ \hline 1\,2500 \\ 5\,000 \\ 25\,00 \\ \hline 31.2500 = \$31.25 \end{array}$$

3. $\frac{\$50}{\$3,300} = 330\overline{)5}$

$= 1.515\%$

$$\begin{array}{r} 1.515\% \text{ monthly} \\ \times 12 \\ \hline 2\,030 \\ 15\,15 \\ \hline 17.180 = 17.18\% \text{ annually} \end{array}$$

$$\begin{array}{r} .01515 \\ 66\overline{)1.00000} \\ -66^{xxx} \\ \hline 340 \\ -330 \\ \hline 100 \\ -66 \end{array}$$

4. $\frac{\$70}{\$4,900} = \frac{1}{70}$

$= 1.429\%$ monthly
$= 17.148\%$ annually

$$\begin{array}{r} .014285 \\ 70\overline{)1.000000} \\ -70^{xxxx} \\ \hline 300 \\ -280 \\ \hline 200 \\ -140 \\ \hline 600 \\ -560 \\ \hline 400 \end{array}$$

$$\begin{array}{r} 1.429\% \text{ monthly} \\ \times 12 \\ \hline 2\,858 \\ 14\,29 \\ \hline 17.148\% \text{ annually} \end{array}$$

5. $7\% - 5\% = 2\%$

$$\begin{array}{r} \$5000 \\ \times .02 \\ \hline \$100.00 \end{array}$$

4 Figuring Your Federal Income Tax

Have you ever heard anyone say, "I can't afford to take on any more work because it would put me in a higher tax bracket"? Have you ever heard anyone say that being in a higher tax bracket means that you actually lose money by working more? The only way that could happen would be if there were a tax bracket of over 100%.

Back during the 1940s, we did have a 90% tax bracket, but never one of 100% or more. In fact, the current maximum tax rate is 39.6%. Today, nearly everyone will be in one of five tax brackets: 0%, 15%, 28%, 31%, or 36%. In general, the poor pay no tax, members of the working class and some members of the middle class pay 15%, and the relatively well-to-do pay more. Exactly how much you pay depends on four factors—your income, the number of dependents you have, your deductions, and the rate of inflation.

If you're really interested in determining your income tax, you can look at my book, *Economics: A Self-Teaching Guide*, 2nd edition (Wiley, 1999), and, of course, for still more detail, you can consult virtually any income tax guide for the current year.

What we're going to do here is work out a few hypothetical income tax problems so that (1) we can demonstrate that if you earn more income you can't end up with less take-home pay and (2) we can show off how well we can calculate percentages.

The federal income tax brackets are raised every year depending on the inflation rate, so we'll make up two sets of hypothetical brackets. Single taxpayers with taxable incomes up to about $25,000 and married couples filing jointly with taxable incomes up to $40,000 are taxed at 15%. Single people earning over $25,000 and couples earning over $40,000 pay 28%.

Here's the kicker. Everyone is entitled to certain deductions and exemptions. Two such deductions, which we'll take up in the next section, are mortgage interest and property taxes. These exemptions and deductions are subtracted from income *before* you calculate your taxes.

To make things interesting, we'll confine our examples to taxpayers whose earnings put them in higher tax brackets.

Problem 1:

Goldspiel just gets by on his salary of $10,000. Because his exemptions and deductions come to exactly $10,000, he pays no federal income tax. But his boss decides to double his salary to $20,000. If Goldspiel's exemptions and deductions remain the same, how much tax does he pay?

Solution:

Total income ($20,000) – deductions and exemptions ($10,000) = taxable income ($10,000). Fifteen percent of $10,000 = $1,500. Note: Goldspiel earns $10,000 more income, pays $1,500 of that to Uncle Sam, and gets to keep the other $8,500. Did Goldspiel's additional earnings put him in a higher tax bracket? They certainly did. He went from the zero bracket to the 15% bracket. Did his take-home pay decline because he went into a higher bracket? No—it rose from $10,000 to $18,500.

Problem 2:

The Wolfheimers used to earn $60,000 between them, but they both got promotions and now earn $80,000. They are entitled to $20,000 in deductions and exemptions. How much tax did they pay before the promotions? How much did they pay after the promotions?

Solution:

Before promotions: Total income ($60,000) – deductions and exemptions ($20,000) = taxable income ($40,000). Fifteen percent of $40,000 = $6,000 taxes paid.

After promotions: Total income ($80,000) – deductions and exemptions ($20,000) = taxable income ($60,000). Fifteen percent of $40,000 = $6,000. Twenty-eight percent of $20,000 = $5,600. Taxes paid = $6,000 + $5,600 = $11,600.

Again, were the Wolfheimers better off or worse off after their promotions? We can see that their higher earnings pushed them from the 15% bracket into the 28% bracket and that their taxes rose from $6,000 to $11,600.

How much take-home pay did they receive before the promotions?

$60,00 – $6,000 = $54,000

How much after?

$80,000 – $11,600 = $68,400

So their after-tax income rose from $54,000 to $68,400 even after the IRS took its cut.

This should demonstrate that your after-tax income will never decrease if your before-tax income rises—even if you end up in a higher tax bracket. If you're still not convinced, just turn any extra earnings over to me. I'll pay the taxes on them and keep what's left over.

SELF-TEST 4

For the following problems, assume the same tax rates and tax brackets that were stated in frame 4.

1. The Krauthammers had a combined taxable income of $90,000 (before deductions and exemptions of $25,000). How much taxes do they pay?

2. Retchnik was earning $50,000 before taxes. He took $20,000 in deductions and exemptions and is a single taxpayer. How much tax does he pay?

ANSWERS

1. $90,000 − 25,000 = $65,000

$40,000	$25,000	$6,000
× .15	× .28	+ 7,000
2 000 00	2 000 00	$13,000 = taxes paid
4 000 0	5 000 0	
$6,000.00	$7,000.00	

2. $50,000 − 20,000 = $30,000

$25,000	$5,000	$3,750
× .15	× .28	+ 1,400
1 250 00	400 00	$5,150 = taxes paid
2 500 0	1 000 0	
$3,750.00	$1,400.00	

5 Mortgage Interest and Taxes

Mortgages are too complex to be discussed very thoroughly in a book of this nature. We can barely begin the job of confusing you, but any banker will be happy to let you know all about fixed and variable rates, adjustable rates, maximum rate changes, prepayments, points, balloon payments, home equity loans, and second mortgages. Many of them have gone to great expense to print mortgage brochures that are virtually incomprehensible.

All we want to do here is stress two things: (1) the role interest plays in your mortgage payments; and (2) the tax advantages of owning rather than renting. After that, it will be up to you to go into your friendly neighborhood bank and make your best deal.

Let's set up a 10.5%, fixed rate, 30-year, $100,000 mortgage. Suppose you wanted to find out four things: (1) your monthly payments; (2) how much money you will pay the bank over the life of the mortgage; (3) how much of this payment will be interest; and (4) how much of your first monthly payment will be interest and how much will be paid on the principal.

Before we even begin, I have a confession to make. You really need a banker and a calculator for question 1. So I've gone ahead and gotten the answer for you. The monthly payments would be $914.74. Using that information, you can, without a calculator, find the answers to questions 2 and 3. Go ahead and make your calculations. I'll wait right here.

What did you get? For the amount of money you will pay the bank over the life of the mortgage, all you need to do is multiply $914.74 by 360, which is the number of months in 30 years. That gives us $329,306.40. To determine how much of that will be interest, just subtract the $100,000 principal and you'll get $229,206.40.

The last question, how much of your first monthly payment will be interest and how much will be principal, was a two-part calculator question, for which I'll provide the answers. The interest came to $863.01, and the payment on the principal was $51.73. What happens, of course, is that because you're paying back such a small proportion of your loan each month, you're paying almost entirely interest during the first few years of repayment. But as you continue to pay off the principal, the interest payments become a smaller and smaller proportion of the monthly payment.

Another important question concerns the amount of your average monthly interest payment over the life of the loan. It is found by dividing total interest payments of $229,200.64 by 360. That gives us $636.68. This number is very important for tax purposes because mortgage interest payments are deductible on your federal income tax, as is property tax. These two considerations usually tip the scales in favor of owning rather than renting.

The large majority of homeowners are in the 28% federal personal income tax bracket. So if they made an additional $1,000 of income, how much additional tax would they pay? Answer: $280. How much is a $1,000 deduction worth, then, to this taxpayer? Answer: $280.

Every deduction you can come up with offsets some taxable income. What owning a home does is give the taxpayer thousands of dollars of income tax deductions. Exactly how many thousands depends on two things: (1) the mortgage interest paid and (2) the local property taxes paid.

Problem 1:

What would be the amount of tax deductions you could claim if you bought a home at the terms we discussed—a $100,000, 30-year mortgage at 10.5%? Assume also a $4,000 deduction for local property taxes.

Solution:

Your deductions would come to $7,640.16 for the mortgage interest ($636.68 × 12) plus $4,000 for the property taxes for a total of $11,640.16.

And how much do those deductions reduce your taxes if you're in the 28% bracket?

$11,640.16 × .28 = $3,259.25

Let's try one more.

Problem 2:

If you were in the 28% bracket and you had monthly mortgage interest payments of $950 and local property taxes of $5,500: (1) how much in tax deductions would you get, and (2) how much would your taxes be reduced?

Solution:

(1) $950 × 12 = $11,400 + $5,500 = $16,900

(2) $16,900 × .28 = $4,732.00

SELF-TEST 5

1. If you are paying a bank a total of $350,000 on a $150,000 mortgage over a 20-year period, how much are your monthly mortgage interest payments? If you are also paying property taxes of $5,000 and are in the 28% federal income tax bracket, how much of a tax deduction do you get for owning a home and by how much do you reduce your taxes?
2. If you are paying a bank a total of $450,000 on a $200,000 mortgage over a 30-year period, how much are your monthly mortgage interest payments? If you are also paying property taxes of $9,000 and are in the 28% federal income tax bracket, how much of a tax deduction do you get for owning a home and how much do you reduce your taxes?

ANSWERS

1. $$\frac{\text{total paid} - \text{loan amount}}{\text{number of payments}} = \frac{\$350{,}000 - \$150{,}000}{12(20)}$$

$$= \frac{\$200{,}000}{240}$$

$$= \frac{\$20{,}000}{24}$$

$$= \frac{\$10{,}000}{12}$$

$$= \frac{\$5{,}000}{6}$$

$$= \$833.33 \text{ monthly payment}$$

Tax deduction:
$\$833{,}33 \times 12 = \$10{,}000 + \$5{,}000 = \$15{,}000$
$\$15{,}000 \times .28 = \$4{,}200$

2. $$\frac{\text{total paid} - \text{loan amount}}{\text{number of payments}} = \frac{\$450{,}000 - \$200{,}000}{12(30)}$$

$$= \frac{\$250{,}000}{360}$$

$$= \frac{\$25{,}000}{36}$$

$$= \frac{\$12{,}500}{18}$$

$$= \frac{\$6{,}250}{9}$$

$$= \$694.44$$

Tax deduction:
$\$694.44 \times 12 = \$8{,}333.28$
$\$8{,}333.28 + \$9{,}000 = \$17{,}333.28$
$\$17{,}333.28 \times .28 = \$4{,}853.32$

23 Business Math

Businesses, like individual consumers, are enmeshed in percentages. Salespeople usually earn commissions, which are specified percentages of sales. Manufacturers offer retailers discounts, which again are calculated on a percentage basis. And finally, there are the profits that are calculated as a percentage of sales and as a percentage of investment.

1 Commissions

Most salespeople work on commission—that is, their earnings depend on their sales. The more they sell, the more they earn.

Commission arrangements vary widely—there are straight commissions, salaries plus commissions, draws against commissions, graduated commissions, salaries plus quota-bonus commissions, and even more exotic combinations.

We'll confine ourselves mainly to straight commissions. Here's an easy one.

Problem 1:
If a saleswoman is paid a 10% commission, how much would she earn on sales of $28,515?

Solution:
$28,515 × .10 = $2,851.50

Problem 2:

How much would she earn in 3 months if her monthly sales were $25,500, $16,300, and $30,400?

Solution:

$25,500
16,300
+30,400
$72,200

$72,200
×.10
$7,220

Problem 3:

How much would a salesman on a 4% commission earn in a year if his quarterly sales were $355,200, $289,710, $216,930, and $401,740?

Solution:

$355,200
289,710
216,930
+401,740
$1,263,580

$1,263,580
×.04
$50,543.20

Problem 4:

A saleswoman is paid 12% commission on her sales to new customers and 8% on sales to regular customers. If she wrote up $6,000 in sales to new customers and $15,000 in sales to regular customers, how much money did she earn in commissions?

Solution:

$6,000
×.12
$720.00

$15,000
×.08
$1,200.00

$720 + $1,200 = $1,920

Let's move into real estate commissions.

Problem 5:

Suppose a real estate agency charges tenants 15% of their annual rent to find them apartments. If that fee is split evenly between the agency and the salesperson, how much does the salesperson make if she finds someone an apartment that costs $500 per month?

Solution:

$500 × 12 = $6,000

```
  $6,000
   ×.075
$450.000
```

Problem 6:

How much would a real estate salesman working for the same company earn by renting someone an apartment that costs $700 per month?

Solution:

$700 × 12 = $8,400

```
  $8,400
   ×.075
  42 000
  588 00
$630.000
```

Problem 7:

Finally, we'll sell a whole house. The agency charges the seller a 4% fee, which it splits evenly with the salesperson. How much does the salesperson earn on the sale of a $150,000 house?

Solution:

```
 $150,000
     ×.02
$3,000.00
```

SELF-TEST 1

Find the commissions earned by each of these salespeople:

1. A 15% commission on sales of $40,000.

2. A 5% commission on sales of $5,500, $4,300, and $6,800.

3. A 10% commission is earned on sales to new customers, and a 5% commission is paid on sales to regular customers. A salesperson wrote up sales of $67,500 to new customers and $137,500 to regular customers.

4. A real estate agency charges tenants 15% of their annual rent to find them apartments. If that fee is split evenly between the agency and the salesperson, how much does the salesperson make if she finds someone an apartment that costs $1,500 per month?

5. If the fee is split evenly between the real estate agent and the agency, how much would a salesperson earn if a $200,000 house is sold and the agency gets a 5% fee?

ANSWERS

1. $\$40,000 \times .15 = \$6,000$

2. $$\begin{array}{r} \$5,500 \\ 4,300 \\ +6,800 \\ \hline \$16,600 \end{array} \qquad \begin{array}{r} \$16,600 \\ \times .05 \\ \hline \$830.00 \end{array}$$

3. $\$67,500 \times .10 = \$6,750$

$$\begin{array}{r} \$137,500 \\ \times .05 \\ \hline \$6,875.00 \end{array} \qquad \begin{array}{r} \$6,750 \\ +6,875 \\ \hline \$13,625 \end{array}$$

4. $\$1,500 \times 12 = \$18,000$

$$\begin{array}{r} \$18,000 \\ \times .075 \\ \hline 90\ 000 \\ 1\ 260\ 00 \\ \hline \$1,350.000 = \$1,350 \end{array}$$

5. $$\begin{array}{r} \$200,000 \\ \times .025 \\ \hline 1\ 000\ 000 \\ 4\ 000\ 00 \\ \hline \$5,000.000 = \$5,000 \end{array}$$

2 Mark-Ups

Mark-ups are just the opposite of mark-downs, which we encountered in frame 1 of chapter 22. Some cynics say that every time an item on sale is marked *down*, it had previously been marked *up*. And they're right! Because the mark-up, usually between 20% and 50% of cost, is the way a business firm is able to cover its overhead and, hopefully, make a profit.

Problem 1:

Suppose you were running an audio store. If it makes sounds, you sell it. Let's say that your overhead—salaries, rent, utilities, advertising, insurance, and everything else—comes to $10,000 a month. Suppose you sold 1,000 units a month. These units cost you $50 each, or $50,000. What would your mark-up be? Ten percent? Twenty percent? More? Less? You tell me.

At 20% you'd just break even:

$\$50,000 \times .20 = \$10,000$

This just covers your overhead. Unless you're in business for your health, you're going to have to have a higher mark-up. How much of a mark-up would you need to show a $5,000 profit?

Solution:

A mark-up of 30%:

$\$50,000 \times .30 = \$15,000$
$\$15,000 - \$10,000(\text{overhead}) = \$5,000$ profit

Problem 2:

How much of a mark-up would you need to show a $20,000 profit?

Solution:

A mark-up of 60%:

$\$50,000 \times .60 = \$30,000$
$\$30,000 - \$10,000(\text{overhead}) = \$20,000$ profit

Needless to say, it is one thing to project a profit and another to make one. The higher your mark-up, the higher your prices, and the lower the number of units sold. That has to be figured in too. But we won't be doing that here. That's economics or business, and this is math.

Problem 3:

An auto dealer has 100 used cars that must be sold this week. He paid $8,000 for each car and wants to cover his overhead of $20,000 and show a profit of at least $20,000. What is the minimum mark-up that he would find acceptable?

Solution:

He needs to sell the cars for at least $40,000 more than he paid for them. He paid $800,000, or $\$8,000 \times 100$.

$$\frac{\$40,000}{\$800,000} = \frac{4}{80} = \frac{1}{20} = 5\%$$

Problem 4:

If he wanted to make a $60,000 profit, what mark-up would he come up with?

Solution:

$$\frac{\$80,000}{\$800,000} = \frac{8}{80} = \frac{1}{10} = 10\%$$

SELF-TEST 2

1. If you have an inventory of 200 units, which cost you $15 per unit, you have an overhead of $5,000, and you want to break even, how much is your mark-up?
2. If you have 500 dresses in your shop for which you paid $40 apiece, you have an overhead of $8,000, and you want to make a profit of $4,000, how much is your mark-up?
3. If you have an inventory of 300 units, which cost you $20 per unit, you have an overhead of $3,000, and you want to make a profit of $5,000, how much is your mark-up?

ANSWERS

1. $200 \times \$15 = \$3{,}000 + \$5{,}000 = \$8{,}000 \qquad \frac{\$5{,}000}{\$3{,}000} = 167\%$

$3000\overline{)5000} = 3\overline{)5.000}$ (quotient 1.666) $= 1.67 = 167\%$

2. $\$40 \times 500 = \$20{,}000 + \$8{,}000 + \$4{,}000 = \$32{,}000 \qquad \frac{\$12{,}000}{\$20{,}000} = \frac{12}{20} = \frac{6}{10} = 60\%$

3. $\$20 \times 300 = \$6{,}000 + \$3{,}000 + \$5{,}000 = \$14{,}000 \qquad \frac{\$8{,}000}{\$6{,}000} = 133\frac{1}{3}\%$

3 Discounting from List Price

There are certain goods that have list prices—books, appliances, cars, and sometimes furniture and clothing. These are also known as sticker prices or manufacturers' suggested retail prices. These prices are set by the manufacturer, who then sells the goods to retailers at a set discount.

In frame 2 we talked about a mark-up charged to the customer by the retailer. Here we have a mark-down given to the retailer by the manufacturer. Let's go over an example.

Problem 1:

If list price on a refrigerator is $500 and the manufacturer gives it to the retailer for 40% off list, how much is the retailer charged?

Solution:

$\$500 \times .60 = \$300$

So far, so good.

Problem 2:

Now the retailer turns around and charges the customer $500, right? Right! So what's the percentage mark-up? (Hint: There's a trick here.)

Solution:

$$\frac{\$200}{\$300} = \frac{2}{3} = 66\frac{2}{3}\%$$

Very interesting. The manufacturer gives the retailer a 40% discount off list price (i.e., a $200 discount off the $500 list). Then the retailer raises the price by the same amount, $200, and charges the customer $500. But the percentage increase is 66 2/3% rather than 40%.

How do you explain that when the price is lowered by $200, it's a 40% decrease, but when it's raised by $200, it's a 66 2/3% increase?

Mathematicians had puzzled over this apparent contradiction for centuries, until finally, around 8000 B.C., they came up with the answer. We're using different bases. When we take $200 off the $500 list price, that's $200/$500 = 40%. But when we go up from $300 to $500, we're putting $200 over a base of $300, $200/$300, and getting 66 2/3%.

To sum up, when we measure percentage changes, we'll find that a given change—$200 in this case—will result in a larger percentage change when we use a smaller base—$300 instead of $500.

Now let's return to the real world.

Do all retailers charge list price? Do I even have to ask? Some charge more, and some, who bill themselves as discount stores, charge less than list. Right now, however, we'll just worry about what a retailer has to pay for the merchandise.

Problem 3:

A lamp store receives a shipment of lamps, which have a list price of $29.95, at a 40% discount. What does the retailer pay for each lamp?

Solution:

$\$29.95 \times .60 = \$17.97$

Problem 4:

A toy store receives a shipment of 100 toys at a 20% discount from their list price of $5.95. How much does the retailer pay for this shipment?

Solution:

$\$5.95 \times .8 = \$4.76 \times 100 = \$476$

Are you getting the hang of it? I certainly hope so, because, naturally, we can't leave well enough alone. The next example gets a bit more complicated.

Problem 5:

An appliance dealer receives a shipment of 100 fans at a 30% discount off list. If she is billed \$2,800, how much is the list price (of one fan)?

Solution:

Let x = the list price

$$\$2{,}800 = 100 \times .7x$$
$$\$28 = .7x$$
$$\frac{\$28}{.7} = \frac{.7x}{.7}$$
$$\$40 = x$$

$$.7\overline{)\$28} = 7\overline{)\$280} \quad \text{quotient: } \$40$$

If you got this one right, go to Self-Test 3. If not, go to the next problem.

Problem 6:

A furniture store receives a shipment of 10 couches at a 20% discount and is billed \$4,792. How much is the list price for one couch?

Solution:

Let x = the list price

$$\$4{,}792 = 10 \times .8x$$
$$\$479.20 = 8x$$
$$\frac{\$479.20}{.8} = \frac{.8x}{.8}$$
$$\$599 = x$$

$$.8\overline{)479.20} = 8\overline{)\$47^{7}9^{7}2} \quad \text{quotient: } \$5\ 9\ 9$$

If you got this one right, then go to Self-Test 3. If you didn't, then maybe it's your algebra. I've probably sent you back there before, but you need to return one more time to frame 1 of chapter 18. If you're still uncertain about finding x, the great unknown, then you might do well to go back to frame 1 of chapter 12, read the entire chapter, and then go to frame 1 of chapter 18. After you've completed chapter 18, go back to Problem 5 in this chapter.

SELF-TEST 3

1. An audio store receives 10 turntables at a 40% discount off their list price of $99.99. How much does the retailer pay for this shipment?
2. A shoe store receives 100 pairs of shoes that list for $49.95 at a 30% discount. How much does the retailer pay for this shipment?
3. An appliance dealer receives 20 electric can openers at a 25% discount off list. If the retailer is billed $299.25, how much is the list price?
4. A video store receives 100 tapes at 35% off list. If the retailer is billed $1,946.75, how much is the list price for one tape?

ANSWERS

1.

$999.90
× .60
$599.9400 = $599.94

2.

$4,995
× .70
$3,496.50

3. Let x = list price

$299.25 = 20 × .75$x$
$299.25 = 15$x$
$19.95 = x

```
      19.95
15 )299.25
   - 15
     149
   - 135
      14 2
    - 13 5
        75
      - 75
```

4. Let x = list price

$1,946.75 = 100 × .65$x$
$1,946.75 = 65$x$
$389.35 = 13$x$
$29.95 = x

```
      29.95
13 )389.35
   - 26
     129
   - 117
      12 3
    - 11 7
        65
      - 65
```

4 Quantity Discounts

Did you hear about the panhandler who asked passersby for 8 cents for a cup of coffee? Finally someone stopped and asked where he could buy a cup of coffee for 8 cents. The panhandler replied, "So who buys retail?"

The whole point of this story is to impress upon you the importance of quantity discounts. Because of shipping and handling costs, most suppliers offer such discounts.

Problem 1:

If Philip Morris offers an 8% discount on deliveries of at least 20 cartons and 12% on deliveries of at least 100 cartons, how much would a store pay for 20 cartons if the regular price charged were $7 a carton (for orders of under 20 cartons)?

Solution:

$\$7 \times 20 = \$140$

$\$140 \times .92 = \$128.80$

Problem 2:

How much would a store pay for an order of 150 cartons?

Solution:

$\$7 \times 150 = \$1,050$

$\$1,050 \times .88 = \$924$

Note the shortcuts we've been taking. Instead of having multiplied \$1,050 by .12 and then subtracting that product from \$1,050, we've saved ourselves a little work. As you keep working with figures, you'll feel comfortable taking these shortcuts too.

Obviously it pays to buy in quantity. But quantity buying does have its costs. You need larger storage facilities, there's sometimes spoilage, there could be obsolescence (especially with fashion-oriented or computer goods), and then there is the money you have tied up in carrying larger inventories. Unless the supplier allows you to buy on credit—and doesn't offer a discount for fast payment—the carrying of a large inventory implies an interest cost for the money you've invested in that inventory.

Problem 3:

A retailer is offered a quantity discount of 5% on all orders of 50 or more and a discount of 8% on orders of 200 or more. How much would an order of 100 cost if the regular price (before the discount) were \$60 per unit?

Solution:

$\$60 \times 100 = \$6,000 \times .95 = \$5,700$

Problem 4:

How much would the retailer have to pay for an order of 300 units?

Solution:

$\$60 \times 300 = \$18,000 \times .92 = \$16,560$

SELF-TEST 4

1. A retailer is offered a quantity discount of 6% on all orders of 100 or more and a discount of 9% on orders of 400 or more. Find how much

the retailer would pay for (a) an order of 100 and (b) an order of 500. Assume a price of $5 per unit before any discounts.

2. A manufacturer offers its retail dealers the following schedule of quantity discounts: on orders over 50, a 2% discount; on orders over 200, a 4% discount; and on orders over 1,000, a 6% discount. How much would a dealer pay for an order of (a) 100, (b) 500, and (c) 2,000? Assume a price of $10 per unit before any discounts.

ANSWERS

1. a. $5 × 100 = $500
   $500 × .94 = $470
   b. $5 × 500 = $2,500
   $2,500 × .91 = $2,275

2. a. $10 × 100 = $1,000
   $1,000 × .98 = $980
   b. $10 × 500 = $5,000
   $5,000 × .96 = $4,800
   c. $10 × 2,000 = $20,000
   $20,000 × .94 = $18,800

5 2/10 n/30

One of the biggest problems some businesses encounter is large backlogs of accounts receivable. Face it folks, there are some firms that are very slow to pay their bills. How slow? Don't ask.

To give their business customers an incentive to pay their bills quickly, many firms offer terms of 2/10 n/30. Literally interpreted, these terms offer the buyer a 2% discount if the bill is paid within 10 days. If the buyer does not choose this option, the full amount must be paid within 30 days. It's n/30, or net/30, because the buyer will also receive a trade discount or discount off list.

So the retailer is given two options: Pay the bill within 10 days and take 2% off, or pay the bill within 30 days and take nothing off. For example, if you paid for a $10,000 shipment within 10 days, you would pay just $9,800, instead of paying the full $10,000 at the end of the month. Which would you do? Of course, there is a third option that many businesses take: They pay late.

Time is money, and, as we've seen in chapter 19, the interest rate is the price charged for the use of money over time. By offering us a 2% discount for fast, or cash, payment, our suppliers are really offering us interest on our money. The question is, how much?

Interest is usually calculated at an annual rate. How much, then, is 2/10 on an annual basis? We'll make two assumptions to simplify this question: (1) you pay in exactly 10 days and (2) the business year consists of 12 30-day months, or 360 days.

Since your bill is due in 30 days, and assuming that you would pay it on the thirtieth day if you did not take the 2% discount and pay on the tenth day, you would have the use of your money for an extra 20 days. So by holding onto your money for 20 days, you are giving up 2% interest. How much does that come to on an annual basis?

You have all the information you need to figure this out. You save 2% on a $10,000 bill by paying it 20 days in advance. The business year has 360 days.

Problem 1:

Work out the annual rate of interest.

Solution:

Let x = the annual rate of interest

We're going to do this step by step. First, we set our basic relationships: 2% interest in 20 days = x% interest in 360 days:

$$2\%:20 = x:360$$
$$2\%(360) = 20x$$

What we've done here is multiply the means (or inside numbers, 20 and x) by the extremes (or outside numbers, 2% and 360). From here on in, it's simple arithmetic:

$$720\% = 20x$$
$$36\% = x$$

This type of problem gives us a chance to show off our algebra from chapter 18 and our ratios and proportions from chapter 14. You might want to glance back at these chapters if you're still uncomfortable with this type of problem.

Incidentally, if you don't think you can find happiness with the method I just used to calculate the annual rate of interest, here's an alternate method.

Problem 2:

If you save 2% by paying 20 days early, how much is the daily interest rate?

Solution:

$$2\% = 20 \text{ days}$$
$$1\% = 10 \text{ days}$$
$$0.1\% = 1 \text{ day}$$

The annual rate of interest is:

$$360 \times 0.1\% = 36\%$$

Which method is better? The one you're more comfortable with.

Problem 3:

What if the terms were 3/10 n/60? How much is the annual rate of interest that you would receive by paying within 10 days? (Hint: Pay within 10 days as opposed to paying how many days later?)

Solutions:

First method: Let x = annual interest rate

$$3\%:50 = x:360$$
$$3\%(360) = 50x$$
$$1{,}080\% = 50x$$
$$108\% = 5x$$
$$21.6\% = x$$

Second method:

$$3\% = 50 \text{ days}$$
$$0.3\% = 5 \text{ days}$$
$$21.6\% = 360 \text{ days}$$

How did we get from the second step to the third step? By multiplying both sides by 72, since this gave us the interest rate for 360 days.

The key thing is to figure out that you either pay in 10 days and get the discount, or pay in 60 days and don't get the discount. Therefore, if you don't pay for 60 days, you have the use of the money for an additional 50 days.

Problem 4:

What if the terms were 2/20 n/60?

Solutions:

First method: Let x = annual interest rate

$$2\%:40 = x:360$$
$$720\% = 40x$$
$$72\% = 4x$$
$$18\% = x$$

Second method:

$$2\% = 40 \text{ days}$$
$$1\% = 20 \text{ days}$$
$$0.1\% = 2 \text{ days}$$
$$0.05\% = 1 \text{ day}$$
$$18\% = 360 \text{ days}$$

Does it make any sense for a business to delay paying its bills? It must make sense to *some* businesspeople. Like crime—if it doesn't pay, how come it's so popular?

If you don't pay your bills promptly, you not only lose the early payer's discount, but credit agencies will list you as a slow payer, and suppliers will be very cautious about granting you credit.

But there's an advantage to slow paying as well. First of all, you might not have the money to begin with. Or, if you do have the money, the longer you delay, the longer you have the use of the money that you owe.

Problem 5:

Terms are 2/10 n/30. The firm finally pays after 4 months. If the bill is for $10,000 and the going rate of interest is 12%, how much is it worth to this retailer to have the use of $10,000 for 120 days?

Solution:

120 days = 1/3 of a year. If the annual rate of interest is 12%, then 1/3 of 12% is 4%.

$$.04 \times \$10{,}000 = \$400$$

It is worth $400 in what we can call implied interest for the retailer to pay 90 days late (or 120 days after being billed). So it does pay to pay late. And the higher the going rate of interest and the longer we delay payment, the more it pays. This may offend your sense of justice, but we've got to let the bad guys win once in a while.

SELF-TEST 5

1. How much is the implied annual interest rate for each of the following payment terms?
   a. 2/10 n/60 b. 1/30 n/90 c. 3/10 n/60
2. On a $1,000 bill, how much does a retailer save by paying it in 6 months if the annual rate of interest is 10%?
3. On a $100,000 bill, how much does a retailer save by paying it 3 months late if the annual rate of interest is 16%?

ANSWERS

1. a. Let x = annual interest rate
   2%:50 = x:360
   720% = 50x
   72% = 5x
   14.4% = x

   b. Let x = annual interest rate
   1%:60 = x:360
   360% = 60x
   6% = x

   c. Let x = annual interest rate
   3%:50 = x:360
   1,080% = 50x
   108% = 5x
   21.6% = x

2. .05 × $1,000 = $50
3. .04 × $100,000 = $4,000

6 Chain Discounts

In this section, we're not talking about mark-downs offered by chain stores. We're talking about multiple discounts offered by suppliers to retail stores. So far we've looked separately at three different discounts. There was the trade discount, or discount from list price. Then there was the quantity discount. And finally, there was the cash discount, which was a reward for prompt payment. A chain discount is a combination of two or three of these discounts.

To calculate a chain discount, we'll start with list price. If a manufacturer offers a trade discount of 20% off list, then the retailer would get an item listed at $100 for $80. A 5% quantity discount would knock this down to $76 ($80 × .95). Finally, we have 2/10 n/30. That would bring us down to $74.48 ($76 × .98).

We've done three consecutive multiplications: $100 × .80; $80 × .95; and $76 × .98. We ended up with $74.48, or a total discount of $25.52. Let's try to calculate our last two discounts in a simpler way. After we take the trade discount of $20, let's take 5% of the $100 list price. That would give us \$5. And then, we'll take 2% of list, or \$2 for our cash discount. \$20 + \$5 + \$2 = \$27.

Not only would that be better for the retailer, but it's easier to figure out. However, we don't do it that way because business convention dictates that we don't. Furthermore, if we carried out this practice to its logical conclusion, we could end up with discounts that totaled more than the list price (see the following box).

Does it matter whether we multiply the trade discount first, then the quantity discount, and finally, the cash discount? In other words, does the order in which we multiply affect our answer? No. If you don't believe me, try it yourself.

Now we'll work out a couple of problems.

Problem 1:

Calculate how much this retailer pays, given these terms: trade discount of 25%, quantity discount of 5%, 2/10 n/30. The list price is \$50.

Solution:

\$50 × .75 × .95 × .98 = \$34.91.

How to Get over 100% Fuel Economy

There are a lot of ways to conserve gasoline. If you keep your speed down to about 35 m.p.h., you may be able to cut your gas consumption by 20% to 25%. If you use the highest octane gas and the finest quality motor oil, you'll save another 30% to 35%. Special oil and air filters add another 10% to fuel economy. Frequent tune-ups add another 20%, and there's a new device that can be installed in your carburetor to save still another 25%.

How much have you saved so far? One hundred and ten percent? That would mean you'd have to stop every 100 miles or so and bail out all that extra gasoline.

Problem 2:

How much does the retailer pay for an order of 500 units if the list price is \$40, the trade discount is 30%, a quantity discount of 4% is offered on orders of over 100, and there are additional terms of 3/10 n/60?

Solution:

$500 \times \$40 \times .7 \times .96 \times .97 = \$13{,}036.80$

SELF-TEST 6

1. How much does the retailer pay for an order of 1,000 units if the list price is $60, the trade discount is 40%, a quantity discount of 5% is offered on orders over 500, and there are additional terms of 2/20 n/60?
2. How much does a retailer pay for an order of 500 units if the list price is $100, the trade discount is 35%, a quantity discount of 4% is offered on orders of over 100, and there are additional terms of 2/10 n/30?
3. How much does a retailer pay for an order of 100 units if the list price is $250, the trade discount is 30%, a quantity discount of 6% is offered on orders of 100 or more, and there are additional terms of 1/20 n/60?

ANSWERS

1. $\$60 \times 1{,}000 \times .60 \times .95 \times .98 = \$33{,}516$
2. $\$100 \times 500 \times .65 \times .96 \times .98 = \$30{,}576$
3. $\$250 \times 100 \times .70 \times .94 \times .99 = \$16{,}285.50$

7 Profit

We've finally gotten down to the bottom line. Profit will be the last new topic that we'll be covering in this book. And, like every other topic in the last two chapters, profit will be calculated as a percentage. But unlike just about everything else we've been calculating, profit is something the businessperson gets to keep.

Economists keep telling us that businesspeople always try to maximize their profits. What we're concerned with here is measuring those profits.

The formula for profit is simple:

sales − costs = profit

Now comes the fun part. Is a profit of $100,000 good or bad? It all depends on what those $100,000 worth of profits are relative to. Are they relative to sales? Or to your investment?

Remember one of the reasons we were given about never shoplifting from supermarkets? Shoplifting was bad, but it was particularly bad to steal from a supermarket. Do you remember being told that if we stole even one item, the supermarket would lose money on us? And if everyone stole just one item, the supermarket would be forced to go out of business because it would be losing money? Don't you remember all that? Maybe I just grew up in a strange neighborhood.

Anyway, the point I'm not very successfully getting at is that supermarkets have extremely low profit margins. That is—they have extremely low profit margins on sales. Maybe they make just one or two cents on every dollar rung up on their cash registers. You see all those people on line at all those checkout counters, and all those purchases represent just a few dollars of profit. Why so few? Because we're talking about profit as a percentage of sales, and even the smallest supermarket does thousands of dollars of sales every day.

The formula for profits as a percentage of sales is:

profit/sales

Now we'll plug in some numbers.

Problem 1:

A firm has sales of \$10,000,000 and costs of \$9,900,000. Find the profit as a percentage of sales.

Solution:

$$\begin{aligned} \text{profit} &= \text{sales} - \text{cost} \\ &= 10{,}000{,}000 - \$9{,}900{,}000 \\ &= \$100{,}000 \end{aligned}$$

$$\begin{aligned} \text{Profit as a percentage of sales} &= \frac{\text{profit}}{\text{sales}} \\ &= \frac{\$100{,}000}{\$10{,}000{,}000} \\ &= \frac{1}{100} \\ &= 1\% \end{aligned}$$

Let's do another one.

Problem 2:

A firm has sales of \$200,000,000 and costs of \$175,000,000. Find the profit as a percentage of sales.

Solution:

$$\begin{aligned} \text{profit} &= \text{sales} - \text{costs} \\ &= \$200{,}000{,}000 - \$175{,}000{,}000 \\ &= \$25{,}000{,}000 \end{aligned}$$

$$\begin{aligned} \text{Profit as a percentage of sales} &= \frac{\text{profit}}{\text{sales}} \\ &= \frac{\$25{,}000{,}000}{\$200{,}000{,}000} \\ &= \frac{25}{200} \\ &= \frac{1}{8} \\ &= 12.5\% \end{aligned}$$

It's nice to know your profit as a percentage of your sales, but some businesspeople find it even nicer to know their profit as a percentage of their investment. After all, why tie up your money in one business if it can earn you a greater return in some other business?

The formula for profit as a percentage of investment should come as no surprise:

profit/investment

How much is your investment? Well, we don't care about your initial investment. The important thing about your investment is what it's worth right now. If you went out of business today and sold your plant and equipment, inventory, your list of customers, and, if possible, your good will, how much could you get for it? That's your investment.

Now we're ready to figure your profit as a percentage of your investment.

Problem 3:

A firm has sales of $10 million, costs of $9 million, and investments of $5 million.

Solution:

$$\begin{aligned} \text{Profit as a percentage of investment} &= \frac{\$1{,}000{,}000}{\$5{,}000{,}000} \\ &= 20\% \end{aligned}$$

Is that good? It's very good. Where else could you earn a 20% return on your capital? But it's not great. Why not? For two reasons. First, you do have your money tied up in your business. Times may be good right now, but who knows if you'll make a 20% profit—or any profit at all—next year or the year after. That's why a lot of people prefer not to tie up much of their money in risky ventures—except, of course, if these people happen to be venture capitalists.

Second, although 20% is very good, if you had your money in New York City real estate back in the late 1970s and early 1980s and were properly leveraged, you would have made at least 50% a year on your investment (see the box titled "Making Money in the Real Estate Market").

We're almost finished. Let's put all of our wisdom together in one problem.

Problem 4:

Find profit as a percentage of sales and as a percentage of investment if sales are \$5 million, costs are \$4.5 million, and investment is \$1 million.

Solution:

$$\begin{aligned} \text{profit} &= \text{sales} - \text{cost} \\ &= \$5{,}000{,}000 - \$4{,}500{,}000 \\ &= \$500{,}000 \end{aligned}$$

$$\begin{aligned} \text{Profit as a percentage of sales} &= \frac{\text{profit}}{\text{sales}} \\ &= \frac{\$500{,}000}{\$5{,}000{,}000} \\ &= \frac{5}{50} = \frac{1}{10} \\ &= 10\% \end{aligned}$$

$$\begin{aligned} \text{Profit as a percentage of investment} &= \frac{\text{profit}}{\text{investment}} \\ &= \frac{\$500{,}000}{\$1{,}000{,}000} \\ &= \frac{5}{10} \\ &= \frac{1}{2} \\ &= 50\% \end{aligned}$$

Making Money in the Real Estate Market

In the late 1970s and the early 1980s, Manhattan real estate prices went through the roof. For example, a building that was purchased for $200,000 in 1977 might have been sold for $800,000 by 1985. If you put down $40,000 cash and obtained a $160,000 mortgage at 10% interest, what was your total profit over this 8-year period? To keep things simple, let's assume a constant interest payment of 10% of the mortgage (i.e., you were not paying back any principal).

$$\text{total profit} = \text{sales} - \text{cost}$$

Your sales would be $800,000. Your cost would be what you paid for the building, $200,000 (your $40,000 plus the $160,000 you paid back the bank in 1977), plus your interest payments of $16,000 a year (10% of $160,000) for 8 years ($\$16{,}000 \times 8 = \$128{,}000$).

$$\begin{aligned}\text{total profit} &= \text{sales} - \text{cost}\\ &= \$800{,}000 - (\$200{,}000 + \$128{,}000)\\ &= \$800{,}000 - \$328{,}000\\ &= \$472{,}000\end{aligned}$$

So your total profit over these 8 years was $472,000. And in 1 year, your profit was $59,000.

Next comes your profit as a percentage of your investment:

$$\frac{\$59{,}000}{\$40{,}000} = \frac{59}{40} = 147.5\%$$

SELF-TEST 7

1. Calculate profit as a percentage of sales:
   a. Sales = $1 million; costs = $950,000
   b. Sales = $500 million; costs = $200 million
2. Calculate profit as a percentage of investment:
   a. Sales = $600,000; costs = $500,000; investment = $1 million
   b. Sales = $50 million; costs = $45 million; investment = $20 million

3. Calculate profit as a percentage of sales and as a percentage of investment:

   a. Sales = \$1 billion; costs = \$950 million; investment = \$400 million

   b. Sales = \$40 million; costs = \$38 million; investment = \$5 million

ANSWERS

1. a. profit = sales − cost
   = \$1,000,000 − \$950,000
   = \$50,000

$$\text{Profit as a percentage of sales} = \frac{\text{profit}}{\text{sales}} = \frac{\$50,000}{\$1,000,000} = \frac{5}{100} = 5\%$$

   b. profit = sales − cost
   = \$500,000,000 − \$200,000,000
   = \$300,000,000

$$\text{Profit as a percentage of sales} = \frac{\text{profit}}{\text{sales}} = \frac{\$300,000,000}{\$500,000,000} = \frac{3}{5} = 60\%$$

2. a. profit = sales − cost
   = \$600,000 − \$500,000
   = \$100,000

$$\text{Profit as a percentage of investment} = \frac{\text{profit}}{\text{investment}} = \frac{\$100,000}{\$1,000,000} = \frac{1}{10} = 10\%$$

   b. profit = sales − cost
   = \$50,000,000 − \$45,000,000
   = \$5,000,000

$$\text{Profit as a percentage of investment} = \frac{\$5,000,000}{\$20,000,000} = \frac{5}{20} = \frac{1}{4} = 25\%$$

3. a. $\text{profit} = \text{sales} - \text{cost}$

$$= \$1{,}000{,}000{,}000 - \$950{,}000{,}000$$
$$= \$50{,}000{,}000$$

$$\text{Profit as a percentage of sales} = \frac{\text{profit}}{\text{sales}}$$
$$= \frac{\$50{,}000{,}000}{\$1{,}000{,}000{,}000}$$
$$= \frac{5}{100} = 5\%$$

$$\text{Profit as a percentage of investment} = \frac{\$50{,}000{,}000}{\$400{,}000{,}000}$$
$$= \frac{5}{40}$$
$$= \frac{1}{8} = 12.5\%$$

b. $\text{profit} = \text{sales} - \text{cost}$

$$= \$40{,}000{,}000 - \$38{,}000{,}000$$
$$= \$2{,}000{,}000$$

$$\text{Profit as a percentage of sales} = \frac{\text{profit}}{\text{sales}}$$
$$= \frac{\$2{,}000{,}000}{\$40{,}000{,}000}$$
$$= \frac{2}{40}$$
$$= \frac{1}{20} = 5\%$$

$$\text{Profit as a percentage of investment} = \frac{\text{profit}}{\text{investment}}$$
$$= \frac{\$2{,}000{,}000}{\$5{,}000{,}000}$$
$$= \frac{2}{5} = 40\%$$

24 Review

Congratulations! You've just begun the last chapter of this book. Absolutely no new material will be introduced here. This chapter is composed of questions drawn from each of the earlier chapters starting with chapter 3. It's kind of a final exam, except that it covers material from maybe 10 years of math rather than from just a term's work.

Almost no one gets 100 on a final exam, so I would be amazed if you got everything right. If you do get a particular problem wrong, you'll have the option of going back to the frames where that problem was introduced and gone over.

Again, don't worry if you get some of these wrong. If you've gotten this far, you've already absorbed about 10 years of math. And if you don't want to go back over each of the problems you get wrong, no one is looking over your shoulder.

Chapter 3 Review

1. $\begin{array}{r} 803 \\ \times 576 \\ \hline \end{array}$

2. $\begin{array}{r} 3195 \\ \times 8917 \\ \hline \end{array}$

3. There are 1,760 yards in a mile. If you drove 67 miles, how many yards did you cover?

4. A car dealer sold 84 cars at $17,095 each. What were her total sales?

Answers for Chapter 3 Review

1.
803
× 576
4818
5621
4015
462528

2.
3195
× 8917
22365
3195
28755
25560
28489815

See frames 1 through 5.

3.
1760
× 67
12320
10560
117920 yards

4.
$17095
× 84
68380
136760
$1,435,980

See frame 6.

Chapter 4 Review

1. $8\overline{)905}$

2. $22\overline{)1371}$

3. $752\overline{)69323}$

4. At a weight reduction center, 135 people lost a total of 20,925 pounds in 2 years. If each person lost exactly the same amount, how many pounds did each person lose?

5. If a path is 756 inches long, how many yards long is that path?

Answers for Chapter 4 Review

1.
1 1 3.1 2 5 = 113.1
$8\overline{)9^{1}0^{2}5.^{1}0^{2}0^{4}0}$

See frame 1.

2.
62.3
$22\overline{)1371.0}$
-132^{xx}
51
−44
7 0
−6 6

3.
92.18 = 92.2
$752\overline{)69323.00}$
-6768^{x}
1643
−1504
139 0
−75 2
63 80
−60 16

See frame 2.

4.
155 pounds
$135\overline{)20925}$
-135^{xx}
742
−675
675
−675

5.
21 yards
$36\overline{)756}$
-72^{x}
36

See frame 3.

Chapter 5 Review

1. Multiply each of these numbers by 10:

a. .008 b. .9 c. 401

2. Multiply each of these numbers by 100:
a. 700 b. 2.1 c. .007

3. Divide each of these numbers by 10:
a. .01 b. 1.3 c. 618

4. Divide each of these numbers by 100:
a. .03 b. 67 c. 4,005

Answers for Chapter 5 Review

1. a. .08
b. 9
c. 4,010

2. a. 70,000
b. 210
c. .7
See frame 1.

3. a. .001
b. .13
c. 61.8

4. a. .0003
b. .67
c. 40.05
See frame 2.

Chapter 6 Review

1. 1.09×8.6

2. 14.562×4.225

3. $8.96\overline{)3.57}$

4. $1.53\overline{)4.02}$

Answers for Chapter 6 Review

1.
```
   1.09
  ×8.6
   654
  8 72
 9.374
```

2.
```
    14.562
   ×4.225
     72810
     29124
    2 9124
   58 248
 61.524450
```
See frame 1.

3.
```
                              .39 = .4
8.96 )3.57 = 896 )357.00
                  - 268 8^x
                     88 20
                   - 80 64
```

4.
```
                   2.6 = 2.6
1.53 )4.02 = 153 )402.00
                    xx
               - 306
                  96 0
                - 91 8
                   4 20
```
See frame 2.

Chapter 7 Review

1. Convert 5/8 into a decimal.

2. Convert 3/4 into a decimal.

3. Convert .6 into a fraction.

4. Convert .45 into a fraction.

Answers for Chapter 7 Review

1. $\frac{5}{8} = 8\overline{)5.0^20^40}$ with quotient $.6\,2\,5$

2. $\frac{3}{4} = 4\overline{)3.0^20}$ with quotient $.7\,5$

See frame 1.

3. $.6 = \frac{.60}{1} = \frac{60}{100} = \frac{6}{10} = \frac{3}{5}$

4. $.45 = \frac{.45}{1} = \frac{45}{100} = \frac{9}{20}$

See frames 2 and 3.

Chapter 8 Review

1. Add these fractions:
a. 1/3 + 1/2 + 3/4
b. 1/5 + 1/2 + 2/3

2. Subtract these fractions:
a. 2/5 − 1/3
b. 3/4 − 2/3

3. Multiply these fractions:
a. 1/4 × 3/5
b. 2/3 × 2/5

4. Find one-eighth divided by one-fourth.

5. Find three-eighths divided by one-seventh.

Answers for Chapter 8 Review

1.

a. $\frac{1\times4}{3\times4} + \frac{1\times6}{2\times6} + \frac{3\times3}{4\times3} = \frac{4}{12} + \frac{6}{12} + \frac{9}{12} = \frac{19}{12} = 1\frac{7}{12}$

Wait—as printed: $= \frac{9}{12}$ then $= 1\frac{7}{12}$

b. $\frac{1\times6}{5\times6} + \frac{1\times15}{2\times15} + \frac{2\times10}{3\times10} = \frac{6}{30} + \frac{15}{30} + \frac{20}{30} = \frac{41}{30} = 1\frac{11}{30}$

See frame 1.

2.

a. $\frac{2\times3}{5\times3} - \frac{1\times5}{3\times5} = \frac{6}{15} - \frac{5}{15} = \frac{1}{15}$

b. $\frac{3\times3}{4\times3} - \frac{2\times4}{3\times4} = \frac{9}{12} - \frac{8}{12} = \frac{1}{12}$

See frame 2.

3.

a. $\frac{1}{4} \times \frac{3}{5} = \frac{3}{20}$

b. $\frac{2}{3} \times \frac{2}{5} = \frac{4}{15}$

See frame 3.

4. $\frac{1}{8} \times \frac{4}{1} = \frac{4}{8} = \frac{1}{2}$

5. $\frac{3}{8} \times \frac{7}{1} = \frac{21}{8} = 2\frac{5}{8}$

See frame 4.

Chapter 9 Review

1. If you walked 17 miles at an average speed of 14 1/4 minutes per mile, how long did it take you to walk the entire distance?

2. How much is three-fifths of two-thirds?

3. If silver is selling at $6.25 an ounce, how much silver could you buy for $580?

Answers for Chapter 9 Review

1. $\frac{17}{1} \times \frac{57}{4} = \frac{969}{4}$

$= 242\frac{1}{4}$ minutes

= 4 hours, 2 minutes, 15 seconds

2. $\frac{2}{3} \times \frac{3}{5} = \frac{6}{15} = \frac{2}{5}$

3. $6.25)$580 = 625)58000
= 125)11600
= 25)2320
= 5)464.0 = 92.8
= 92.8 ounces

Chapter 10 Review

1. Convert 17/100 into a percent.

2. Convert .82 into a percent.

3. Convert 5/8 into a percent.

4. If you grew from 5 feet 4 inches to 5 feet 9 inches, by what percent did your height increase?

5. A change from 15 to 60 represents a percentage change of how much?

6. If a high school has 100 freshmen, 90 sophomores, 80 juniors, and 70 seniors, find the percentage share of the student body of each class.

7. If you left a 15% tip, how much would you leave on a restaurant check of $44.80?

On Mistakes

Some people are prone to making mistakes. They add up a column of figures, and half the time it comes out wrong. Their checkbooks never balance because they forgot to enter every check and every deposit.

This does not mean these people are bad citizens, nor does it even mean they're bad at math. And it certainly doesn't mean they can't do the problems in this book. So don't worry about it if every problem doesn't come out right.

Back in the army we used to have a saying—Don't sweat the small stuff. Actually we didn't use the word "stuff," but our meaning was clear. There are enough big things in life to worry about without losing sleep over every minor thing that goes wrong.

So if you're not getting the right answer to every single problem, you're forgiven. Everyone is going to make some numerical errors. It's only human. Of course, you don't want to overdo your humanity.

Answers for Chapter 10 Review

1. 17%

2. 82%

3. $8\overline{)5.0^{2}0^{4}0}$ = .6 2 5 = 62.5%

See frame 1.

4. $\frac{5 \text{ inches}}{64 \text{ inches}}$

64)5.000 = .078 = 7.8%
−4 48ˣ
520
−512
8

See frame 2.

5. 300%

See frame 3.

6.

freshmen	100	29.4%
sophomores	90	26.5
juniors	80	23.5
seniors	70	20.6
	340	100.0%

freshmen: $\frac{100}{340} = \frac{10}{34} = \frac{5}{17}$

17)5.000 = .294 = 29.4%
−3 4ˣˣˣ
1 60
−1 53
70

sophomores: $\frac{90}{340} = \frac{9}{34}$

34)9.000 = .264 = 26.5%
−6 8ˣˣ
2 20
−2 04
160
−136
24

juniors: $\frac{80}{340} = \frac{8}{34} = \frac{4}{17}$

17)4.000 = .235 = 23.5%
−3 4ˣˣ
60
−51
90
−85
5

seniors: $\frac{70}{340} = \frac{7}{34}$

34)7.000 = .206 = 20.6%
−6 8ˣˣ
200
−204

See frame 4.

7. Ten percent of $44.80 is $4.48, and one-half of $4.48 is $2.24. Your tip would be $6.72 ($4.48 + $2.24).
See frame 5.

Chapter 11 Review

1. Add -4 and -5.

2. Add $-3, +4, -1$, and -2.

3. Multiply -2 and -3.

4. Multiply 5 and -2.

5. Divide -8 by -2.

6. Divide 9 by -3.

Answers for Chapter 11 Review

1. -9

2. -2
See frames 1 through 3.

3. $+6$

4. -10
See frame 4.

5. $+4$

6. -3
See frame 5.

Chapter 12 Review

1. Find x:
a. $x+4=9$
b. $x-3=8$

2. Find x:
a. $3x=20$
b. $\frac{x}{5}=1$

3. Find x:
a. $.2x=3$
b. $1.5x=2$

4. Find x:
a. $x+2=-5$
b. $x+4=3$

5. Find x:
a. $5x-2=8$
b. $\frac{3}{4}x+5=8$
c. $4.2x+7=1$

Answers for Chapter 12 Review

1. a. $x=5$
b. $x=11$
See frames 1 through 3.

2. a. $x=6\frac{2}{3}$
b. $x=5$
See frame 4.

3. a. $x=15$
b. $x=2/1.5$
$=4/3$
$=1\ 1/3$
See frame 5.

4. a. $x = -7$
b. $x = -1$
See frame 6.

5. a. $x = 2$
b. $3/4x = 3$
$x = 4$
c. $4.2x = -6$
$42x = -60$
$21x = -30$
$7x = -10$
$x = -1\ 3/7$
See frame 7.

Chapter 13 Review

1. If $x = 4$, how much is x^1?

2. If $x = 5$, how much is x^3?

3. If $x = 2$, how much is x^4?

4. What is the square root of 81?

5. What is the square root of 49?

6. How much is x, if $x^2 = 16$?

7. How much is x, if $x^2 = 100$?

8. How much is x, if $2x^2 - 2 = 16$?

9. How much is x, if $3x^2 + 5 = 53$?

10. If $x = 9$, how much is $160 - x^2 + 3x$?

11. If $x = 8$, how much is $4x^2 + 3x - 190$?

Answers for Chapter 13 Review

1. 4

2. $5 \times 5 \times 5 = 125$

3. $2 \times 2 \times 2 \times 2 = 16$
See frame 1.

4. 9

5. 7

6. $x = \pm 4$

7. $x = \pm 10$
See frame 2.

8. $2x^2 = 18$
$x^2 = 9$
$x = \pm 3$

9. $3x^2 = 48$
$x^2 = 16$
$x = \pm 4$

10. $160 - 81 + 27 = 106$

11. $(4 \times 64) + 24 - 190 = 90$

    See frame 3.

Chapter 14 Review

1. What is the ratio of inches per foot?

2. If an employee was out sick on 4 of 72 work days, what is her ratio of sick days to days worked?

3. Solve for x: 8 is to 5 as 40 is to what?

4. $x{:}5 = 12{:}20$. Find x.

5. If you can drive 300 miles in 6 1/4 hours, how long would it take you to drive 700 miles at the same rate?

Answers for Chapter 14 Review

1. 12:1

2. 1:17

See frame 1.

3. $8{:}5 = 40{:}x$
$8x = 200$
$x = 25$

4. $20x = 60$
$x = 3$

5. $300{:}\frac{25}{4} = 700{:}x$
$300x = 25 \times 175$
$300x = 4{,}375$
$x = 14.58$ hours

$$300\overline{)4375} = 3\overline{)4^{1}3.17^{2}5} \quad (14.58)$$

See frame 2.

Chapter 15 Review

1. What is the area of a rectangle that is 8 feet long and 5 feet wide?

2. What is the area of a square that has a side of 5 feet?

3. If land is selling for $200 a square foot, how much would you have to pay for a rectangular lot that is 200 feet by 40 feet?

4. If a room were 20 feet by 45 feet and carpeting cost $20 per yard, how much would it cost for wall-to-wall carpeting?

5. How much would it cost to put a fence around a field that is 40 feet by 150 feet if fencing cost $15 a foot?

6. Find the perimeter of a square lot whose side is 20 feet.

7. If a triangle has a base of 10 inches and is 6 inches high, what is its area?

Answers for Chapter 15 Review

1. 40 square feet **2.** 25 square feet
See frame 1. **3.** 8,000 square feet $\times$ \$200 = \$1,600,000

4. 900 square feet = 100 square yards $\times$ \$20 = \$2,000
See frame 2.

5. 80 feet + 300 feet = 380 feet $\times$ \$15 = \$5,700

6. 80 feet
See frame 3.

7. 30 square inches
See frame 4.

Chapter 16 Review

1. If the diameter of a circle is 10 inches, how much is its circumference?

2. If the circumference of a circle is 15 feet, how much is its diameter?

3. If the radius of a circle is 5 inches, how much is its area?

4. If the diameter of a circle is 10 feet, how much is its area?

Answers for Chapter 16 Review

1. $\frac{220}{7}$ inches $= 31\frac{3}{7}$ inches

2. 15 feet $= \pi D$

15 feet $= \frac{22}{7} \times D$

105 feet $= 22D$

4.8 feet $= D$

$$\begin{array}{r} 4.77 \\ 22\overline{)105.00} \\ -88^{xx} \\ \hline 17\,0 \\ -15\,4 \\ \hline 1\,60 \end{array}$$

See frame 1.

3. $25 \times \frac{22}{7} = \frac{550}{7}$

$= 78\frac{4}{7}$ square inches

$$\begin{array}{r} 7\,8\frac{4}{7} \\ 7\overline{)55^{6}0^{7}} \end{array}$$

4. $78\frac{4}{7}$ square feet

See frame 2.

Chapter 17 Review

1. If a plane took off at noon and flew at 500 m.p.h. until 4:30 P.M., how far did it fly?

2. Joan left work at 5 P.M. and walked 4 m.p.h. until 6:30 P.M. and then took a bus the rest of the way home. How far does she live from work if the bus traveled at 10 m.p.h. and left her at her door at 7:00 P.M.?

3. If a plane went 3,000 miles in 5 1/2 hours, what was its average rate of speed?

4. If Mike drove from home to work, a distance of 40 miles, in 45 minutes and returned by a different route in 1 hour, what was his average rate of speed?

5. Two trains left the station traveling in opposite directions. One train was traveling at the rate of 60 m.p.h., and the other was traveling at the rate of 55 m.p.h. When they were 460 miles apart, for how long had they been traveling?

6. Marcia walked at an average rate of speed of 3 1/2 m.p.h. and covered a 2 distance of 16 miles. How long did she walk?

7. How much time would you save by driving for 100 miles at 65 m.p.h. rather than 55 m.p.h.?

8. How much farther would you get if you drove for 30 minutes at 65 m.p.h. rather than 55 m.p.h.?

Answers for Chapter 17 Review

1. $4.5 \times 500 = 2{,}250$ miles

2. $6 + 5 = 11$ miles
   See frames 1 through 3.

3. $\frac{3{,}000}{5.5} = \frac{6{,}000}{11} = 545.5$ m.p.h.

4. $\frac{80}{1\frac{3}{4}} = \frac{80}{1} \div \frac{7}{4}$
   $= \frac{80}{1} \times \frac{4}{7}$
   $= \frac{320}{7}$
   $= 45\frac{5}{7}$ m.p.h.
   See frame 4.

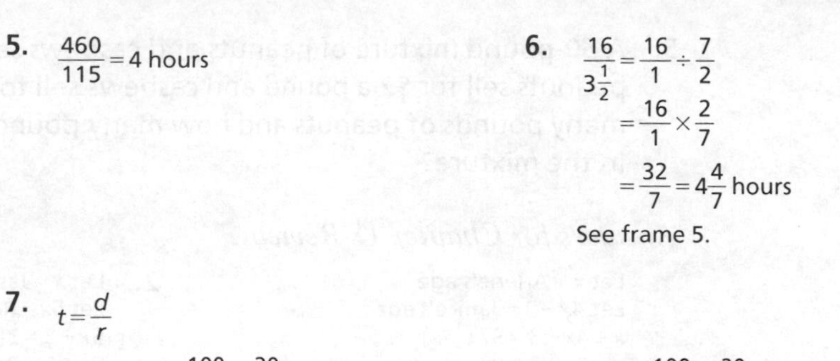

5. $\frac{460}{115} = 4$ hours

6. $\frac{16}{3\frac{1}{2}} = \frac{16}{1} \div \frac{7}{2}$

$= \frac{16}{1} \times \frac{2}{7}$

$= \frac{32}{7} = 4\frac{4}{7}$ hours

See frame 5.

7. $t = \frac{d}{r}$

55 m.p.h. $= \frac{100}{55} = \frac{20}{11}$

```
       1.818 = 1.818 hours
 11 )20.000
   − 11 xxx
      9 0
    − 8 8
       20
     − 11
       90
```

65 m.p.h. $= \frac{100}{65} = \frac{20}{13}$

```
       1.538 = 1.538 hours
 13 )20.000
   − 13 xxx
      7 0
    − 6 5
       50
     − 39
      110
    − 104
        6
```

1.818 hours
− 1.538 hours
$.280 \times 60$ minutes $= 16.800$

$t = 16.8$ minutes

8. $d = r \times t$

$55 = 55 \times \frac{1}{2} = 27\frac{1}{2}$ miles

$65 = 65 \times \frac{1}{2} = 32\frac{1}{2}$ miles

$d = 32\frac{1}{2} - 27\frac{1}{2} = 5$ miles

See frames 6 through 7.

Chapter 18 Review

1. If Janice is 3 years short of being four times Arlene's age, and the sum of their ages is 97, how old are they?
2. Ron is five times Jason's age. In 5 years he will be just three times as old as Jason. How old are they?
3. Find three consecutive numbers adding up to 51.
4. Find three numbers adding up to 72, if the second number is three times as large as the first and the third is 2 larger than the second.

5. A 50-pound mixture of peanuts and cashews sells for \$6 a pound. If peanuts sell for \$2 a pound and cashews sell for \$8 a pound, how many pounds of peanuts and how many pounds of cashews are used in the mixture?

Answers for Chapter 18 Review

1. Let x = Arlene's age
Let $4x-3$ = Janice's age
$x+4x-3=97$
$5x-3=97$
$5x=100$
$x=20$
$4x-3=77$

2. Let x = Jason's age now
Let $5x$ = Ron's age now
$5x+5=3(x+5)$
$5x+5=3x+15$
$5x=3x+10$
$2x=10$
$x=5$
$5x=25$

See frame 1.

3. Let x = first number
Let $x+1$ = second number
Let $x+2$ = third number
$3x+3=51$
$3x=48$
$x=16$
$x+1=17$
$x+2=18$

4. Let x = first number
Let $3x$ = second number
Let $3x+2$ = third number
$7x+2=72$
$7x=70$
$x=10$
$3x=30$
$3x+2=32$

See frame 2.

5. Let x = pounds of peanuts
Let $50-x$ = pounds of cashews
value of peanuts = $\$2x$
value of cashews = $\$8(50-x) = \$400 - \$8x$
value of the entire mixture = $\$6 \times 50 = \$300$
$\$2x + \$400 - \$8x = \$300$
$\$400 - \$6x = \$300$
$-\$6x = -\$100$
$\$6x = \$100$
$x = 16\frac{2}{3}$ pounds
$50 - x = 33\frac{1}{3}$ pounds

See frame 3.

Chapter 19 Review

1. Calculate the simple rate of interest paid on \$1,000 for 6 months at an annual rate of 6%.

2. Calculate the simple rate of interest paid on \$2,000 for 2 years at an annual rate of 7%.

3. Ten thousand dollars is put in a bank that pays interest at an annual rate of 12%, compounded quarterly. How much money would be in the account after 3 quarters?

4. Four thousand dollars is lent out at an annual rate of 8% for 1 year. If the interest is compounded quarterly, how much money would the borrower owe the lender after 1 year?

5. How much is the doubling time for a compound annual interest rate of 7%?

6. How much is the doubling time for a compound annual interest rate of 1%?

7. What is the actual interest rate for each of the following two bank discounts? Each loan is for 1 year:
   a. A $20,000 loan discounted at 9%
   b. A $100,000 loan discounted at 12%

8. Find the true rate of interest for each of these:
   a. A $10,000 bank loan, discounted at 12%, with quarterly payments for 1 year
   b. A $50,000 bank loan, discounted at 8%, with monthly payments for 1 year

Answers for Chapter 19 Review

1. $\$1,000 \times .03 = \$30$

2. $\$2,000 \times 2(.07) = \$2,000 \times .14 = \$280$
See frame 1.

A Word on Formulas

Can you show where it is written that you must memorize a lot of formulas? In chapter 19 we happen to have covered a few fairly complex formulas. I give you my word that I will not think that you are a bad person if you look up these and other formulas.

Formulas are tools to be used, not memorized. The only thing you need to memorize is the multiplication table up to 10×10.

---

3. $A = P(1+r)^t$
$= \$10{,}000(1+.03)^3$
$= \$10{,}000(1.03)^3$
$= \$10{,}000 \times 1.092727$
$= \$10{,}927.27$

```
  1.03        10609
× 1.03       × 1.03
   309        31827
 1030       1 06090
 10609      1.092727
```

4. $A = P(1+r)^t$
$= \$4{,}000(1+.02)^4$
$= \$4{,}000(1.02)^4$
$= \$4{,}000 \times 1.08243216$
$= \$4{,}329.73$

```
  1.02        1.0404       1.08243216
× 1.02      × 1.0404          × 4000
   204         41616       4329.72864
 1 020        416160
 1.0404     1 04040
            1.08243216
```

See frame 2.

5. 10 years

6. 70 years
See frame 3.

7. a. $\frac{\$1{,}800}{\$18{,}200} = \frac{18}{182} = \frac{9}{91}$

```
     .0989 = 9.9%
91 )9.0000
  − 8 19xx
      810
    − 728
      820
    − 819
```

b. $\frac{\$12{,}000}{\$88{,}000} = \frac{12}{88} = \frac{3}{22}$

See frame 4.

```
     .136 = 13.6%
22 )3.000
   − 2 2xx
      80
    − 66
     140
   − 132
       8
```

8. a. $i = \frac{2MC}{P(N+1)} = \frac{2 \times 4 \times \$1{,}200}{\$8{,}800(4+1)}$

$= \frac{8 \times \$1{,}200}{\$8{,}800(5)} = \frac{1 \times \$240}{\$1{,}100 \times 1}$

$= \frac{24}{110}$

$= \frac{12}{55}$

$= 21.8\%$

```
     .218
55 )12.000
   − 11 0xx
      1 00
     − 55
       450
     − 440
        10
```

b. $i = \frac{2MC}{P(N+1)} = \frac{2 \times 12 \times \$4,000}{\$46,000(12+1)}$
$= \frac{24 \times \$4,000}{\$46,000 \times 13} = \frac{24 \times 4}{46 \times 13}$
$= \frac{12 \times 4}{23 \times 13}$
$= \frac{48}{299}$
$= 16.1\%$

$$\begin{array}{r} .1605 \\ 299\,\overline{)48.0000} \\ -29\,9^{xxx} \\ \hline 18\,10 \\ -17\,94 \\ \hline 1600 \\ -1495 \\ \hline 105 \end{array}$$

See frame 5.

Chapter 20 Review

1. Translate these words into numbers:
  a. seven hundred thirty-six thousand, two hundred
  b. forty-three million
  c. five hundred million and sixteen

2. Please express these numbers in words:
  a. 987,100,000
  b. 189,000
  c. 1,000,000

3. Translate these words into numbers:
  a. four trillion
  b. two billion
  c. six hundred forty-one trillion, five hundred billion, seven hundred forty-eight thousand, two hundred and sixty four

4. Please express these numbers in words:
  a. 7,000,000,000,000
  b. 983,000,000,000
  c. 534,987,003,844,010

5. How much is a thousand billion?

6. How much is $100,000 \times 87,661$?

7. How much is $4,130 \times 200,000$?

8. Divide 890 million by 1,000.

9. Divide 450 billion by 10,000.

10. What is the per-capita GDP of a nation that has a population of 50 million and a GDP of $200 billion?

Answers for Chapter 20 Review

1. a. 736,200
   b. 43,000,000
   c. 500,000,016

2. a. nine hundred eighty-seven million, one hundred thousand
   b. one hundred eighty-nine thousand
   c. one million

   See frame 1.

3. a. 4,000,000,000,000
   b. 2,000,000,000
   c. 641,500,000,748,264 (I know, that was a trick question, because I left out millions. So, if you got it wrong, don't lose any sleep over it.)

4. a. seven trillion
   b. nine hundred eighty-three billion
   c. five hundred thirty-four trillion, nine hundred eighty-seven billion, three million, eight hundred forty-four thousand, and ten

   See frame 2.

5. one trillion, or, 1,000,000,000,000

6. 8,766,100,000

7. 826,000,000

8. 890,000

9. 45,000,000

10. $$\text{per-capita GDP} = \frac{\text{GDP}}{\text{population}}$$
    $$= \frac{\$200{,}000{,}000{,}000}{50{,}000{,}000}$$
    $$= \$4{,}000$$

    See frames 3 through 5.

Chapter 21 Review

1. Find the mean, median, and mode of this array: 19,1,8,1,5,6,9

2. Find the mean, median, and modes of this array: 2,23,7,12,6,2,7,5

Answers for Chapter 21 Review

1. mean = 7, median = 6, mode = 1

2. mean = 8, median = 6.5, modes = 2,7

   See frame 1.

Chapter 22 Review

1. A living room set is marked down from $969 to $629. By what percent has the price been cut?

2. An auto dealer is offering a $1,500 rebate on a $20,000 car. What percentage of the original price do you get back?

3. If the sales tax on a $600 purchase is 5%, how much tax would you pay?

4. How much would the original price be if you paid a total (including taxes) of $428 for a sofa and the sales tax were 7%?

5. How much interest would you pay on a credit card balance of $2,500 in 1 month if the annual rate of interest were 16%?

6. If you owe $50 interest on a monthly credit card balance of $4,000, what are the annual and monthly interest rates that you must pay?

7. Jones was earning $60,000 before taxes. He took $20,000 in deductions and exemptions and is a single taxpayer in the 28% tax bracket. The next year he earned $70,000 and took the same exemptions and deductions. How much more tax does he pay this year?

8. You paid the bank a total of $500,000 on a $200,000 mortgage over a 30-year period. You also pay property taxes of $8,000 and are in the 28% federal income tax bracket.
   a. How much are your monthly mortgage payments?
   b. How much of a tax deduction do you get for owning a home?
   c. By how much do you reduce your taxes?

Answers for Chapter 22 Review

1. $\frac{340}{969} = 35.1\%$

$$\begin{array}{r} .351 \\ 969\overline{)340.000} \\ -290\ 7^{xx} \\ \hline 49\ 30 \\ -48\ 45 \\ \hline 850 \end{array}$$

2. $\frac{\$1,500}{\$20,000} = 7.5\%$

See frame 1.

3. $\$600 \times .05 = \$30$

4. Let x = original price

$1.07x = \$428$

$x = \dfrac{\$428}{1.07}$

$x = \$400$

See frame 2.

5. $\dfrac{16}{12} = 1\dfrac{4}{12} = 1.33\%$

$= 2500$

$$\begin{array}{r} 2500 \\ \times .0133 \\ \hline 7500 \\ 7\ 500 \\ 25\ 00 \\ \hline \$33.2500 = \$33.25 \end{array}$$

6. $\dfrac{\$50}{\$4000} = \dfrac{5}{400}$

$= \dfrac{1}{80}$

$= 1.25\%$ monthly

$$\begin{array}{r} 1.25 \\ \times 12 \\ \hline 2\ 50 \\ 12\ 5 \\ \hline 15.00\% \text{ annually} \end{array}$$

See frame 3.

7. $\$10{,}000 \times .28 = \$2{,}800$

See frame 4.

8. a. $\dfrac{\$500{,}000}{360} = \dfrac{50000}{36}$

$= \dfrac{25000}{18}$

$= \dfrac{12500}{9}$

$$9\,\overline{)\,12^{3}5^{8}0^{8}0.^{8}0^{8}0^{8}0} = 1388.888$$

monthly payments = $1,388.89

b. $\dfrac{\$300{,}000}{360} = \dfrac{30000}{36}$

$= \dfrac{10000}{12}$

$= \dfrac{5000}{6}$

$= \$833.33$

$$\begin{array}{r} \$833.33 \\ \times 12 \\ \hline 1\ 666\ 66 \\ 8\ 333\ 3 \\ \hline \$9{,}999.96 \end{array}$$

$$\begin{array}{r} = \$10{,}000 \\ +8{,}000 \\ \hline \$18{,}000 \text{ tax deduction} \end{array}$$

c.

$$\begin{array}{r} \$18{,}000 \\ \times .28 \\ \hline 1\ 440\ 00 \\ 5600 \\ \hline \$7{,}040.00 \text{ tax reduction} \end{array}$$

See frame 5.

Chapter 23 Review

1. A 10% commission is earned on sales to new customers, and a 4% commission is paid on sales to regular customers. A saleswoman wrote up sales of $74,300 to new customers and $110,000 to regular customers. How much did she earn in commissions?

2. A real estate agency charges prospective tenants 15% of their annual rent to find them apartments. If that fee is split evenly between the agency and the salesperson, how much does the salesperson make if she finds someone a $900 apartment?

3. If you have an inventory of 100 units, which cost you $80 per unit, you have an overhead of $4,000. If you want to make a profit of $2,000, how much is your mark-up?

4. A shoe store receives 100 pairs of shoes, which list for $59.95, at a 40% discount. How much does the owner pay for this shipment?

5. A manufacturer offers its retail dealers the following schedule of quantity discounts: on orders over 100, a 3% discount; on orders over 500, a 5% discount; and on orders over 1,000, a 7% discount. If the list price is $20, how much would a dealer pay for the following orders?
a. 150
b. 1,500

6. How much is the implied annual interest rate for these two terms?
a. 2/10 n/30
b. 3/20 n/60

7. How much does a retailer pay for an order of 1,000 units if the list price is $40, the trade discount is 40%, a quantity discount of 5% is offered on orders of 500 or more, and there are additional terms of 2/10 n/60?

8. Calculate profit as a percentage of sales and investment when sales = $1 billion, cost = $900 million, and investment is $400 million.

Answers for Chapter 23 Review

1. $7,430 + $4,400 = $11,830

2.

$$\begin{array}{r} \$900 \\ \times 12 \\ \hline 1800 \\ 900 \\ \hline \$10800 \\ \times .075 \\ \hline 54000 \\ 75600 \\ \hline \$810.000 \end{array}$$

See frame 1.

3. Cost $= \$80 \times 100 = \$8{,}000$

$$\frac{\$6{,}000}{\$8{,}000} = \frac{3}{4} = 75\%$$

See frame 2.

4.

$$\begin{array}{r} \$59.95 \\ \times .6 \\ \hline \$35.970 \times 100 = \$3{,}597 \end{array}$$

See frame 3.

5. a. $150 \times \$20 = \$3{,}000$

$$\begin{array}{r} .97 \\ \times \$3000 \\ \hline \$2{,}910.00 \end{array}$$

b. $1{,}500 \times \$20 = \$30{,}000$

$$\begin{array}{r} .93 \\ \times \$30000 \\ \hline \$27{,}900.00 \end{array}$$

See frame 4.

6. a. Let x = annual rate of interest

$$\begin{aligned} 2\%:20 &= x:360 \\ 720\% &= 20x \\ 72\% &= 2x \\ 36\% &= x \end{aligned}$$

b. Let x = annual rate of interest

$$\begin{aligned} 3\%:40 &= x:360 \\ 1{,}080\% &= 40x \\ 108\% &= 4x \\ 27\% &= x \end{aligned}$$

See frame 5.

7. $1{,}000 \times \$40 \times .6 \times .95 \times .98 =$
$\$24{,}000 \times .931 = \$22{,}344$

See frame 6.

8. profit = sales − cost
$= \$1{,}000{,}000{,}000 - \$9000{,}000{,}000 = \$100{,}000{,}000$

$$\text{Profit as a percentage of sales} = \frac{\text{profit}}{\text{sales}} = \frac{\$100{,}000{,}000}{\$1{,}000{,}000{,}000{,}000} = \frac{1}{10} = 10\%$$

$$\text{Profit as a percentage of investment} = \frac{\text{profit}}{\text{investment}} = \frac{\$100{,}000{,}000}{\$400{,}000{,}000} = \frac{1}{4} = 25\%$$

See frame 7.

Where To from Here?

For most readers, enough is enough! You've relearned most of the math you've forgotten since you got out of school. And as I've said, if you don't use it, you'll lose it.

So try to keep using as much as you can of what we've covered here in your daily life. And stay away from your calculator unless you really need to use it (see chapter 1).

But suppose that you'd like to apply your newly reacquired skills to the study of more advanced math. What I would suggest is that you begin with algebra. We've covered a good bit of algebra in this book, but you can build on that and keep going. Two books I'd recommend are *Quick Algebra Review*, 2nd edition (Wiley) and *Practical Algebra*, 2nd edition (Wiley), both by Peter Selby and Steve Slavin. A more detailed book that does a good job is Douglas Downing's *Algebra the Easy Way* (Barrons).

You might be interested in moving in a somewhat different direction. We've had just a smattering of statistics (in chapter 21), a field that has a wealth of practical applications. If you're interested in this field, the books I'd recommend would be Donald Koosis's *Statistics*, 4th edition (Wiley), and *Chances Are: The Only Statistics Book You'll Ever Need* (University Press of America). If you'd like to continue studying business math, which we covered in chapter 23, see *Quick Business Math* (Wiley) by Steve Slavin.

Another way you might want to go is to take a formal course in either algebra or statistics. Every college offers these courses, and they are almost always available in adult education programs.

Once again, I want to congratulate you on having finished this book. I'm not sure the book ended the way you expected it to, but I think the good guys did win. And even now you're probably asking yourself who was that masked man on the white horse, and why did he leave me with a silver bullet?

Index

addition
fractions, 41–44, 209
isolation of x, the unknown, 73–74
multiplication as, 4
negative numbers, 68, 69, 212
age problems, 131–34, 217–18
algebra
further study of, 227
isolation of x, the unknown, 72–82, 212–13
addition and subtraction, 73–74
combination problems, 79–82
decimals, 76–77
multiplication and division, 74–76
negative numbers, 77–79
areas
of circles, 108–11
of rectangles, 95–100, 214–15
of triangles, 102–4
average, 163–65, 222

big numbers
billions, 155–58, 221–22
dividing, 160–62, 221–22
millions, 153–55, 221–22
multiplying, 159–60, 221–22
quadrillions, 158–59
quintillions, 158–59
thousands, 153–55, 221–22
trillions, 155–58, 221–22
billions, 155–58
business math, 183–205

calculator
dependency on, 2
square roots, 85
uses for, 2
cancelling out, 49–50
Carmen, Marilyn J., 2
Carmen, Robert A., 2
cash discounts, 193–97, 225–26
chain discounts, 197–99, 225–26
circles
circumferences, 105–8, 215
diameters, 105–8, 215
radii, 108–11, 215
commissions, 183–86, 225–26
consecutive number problems, 134–37, 217–18
credit card problems, 173–76, 223–24

decimals
and algebra, 76–77
conversion into fraction, 38–39, 208–9
conversion into percentage, 56
and division, 23–25, 30–31, 34–35
and multiplication, 29–30, 32–34, 207–8
discounting, 148–49
discounting from list price, 188–91, 225–26
distance = rate × time problems, 112–15, 119–20, 125–30, 216–17
division
of big numbers, 160–62
with decimals, 23–25, 30–31, 34–35
fast, 30–31
fractions, 47–48, 209
isolation of x, the unknown, 74–76
long, 22–25, 207
negative numbers, 71, 212
short, 19–22, 207
doubling time, 146–48
Downing, Douglas, 227

equations, 72

exponents, 83–85, 87–88, 213–14

federal income tax calculation, 176–79, 225–26
fractions
  adding, 41–44, 209
  conversion into decimals, 36–37
  conversion into percentages, 56–59, 210
  dividing, 47–48, 209
  lowest common denominator, 39–40, 42–44, 51–53
  multiplying, 45–48, 209
  subtracting, 44–45, 209

interest rate problems
  compound, 143–46, 218–20
  discounting, 148–49, 218–20
  doubling time, 146–48
  mortgage, 176–79
  simple, 141–43, 218–20
  true, 150–52, 218–20

Koosis, Donald, 227

lowest denominators of fractions, 39–40, 42–44, 51–53

mark-down problems, 166–70, 223
mark-ups, 186–88, 225–26
mean, 163–65, 222
median, 163–65, 222
millions, 153–55, 221–22
mode, 163–65, 222
mortgage interest, 176–79, 225–26
multiple discounts, 197–99, 225–26
multiplication
  as addition, 4
  big numbers, 159–60
  with decimals, 29–30, 32–34, 207–8
  fast, 28–30
  fractions, 45–48, 209
  isolation of *x*, the unknown, 74–76
  long, 10–18, 206–7
  negative numbers, 69–71, 212
  simple, 4–9
  table, 2, 5–9

negative numbers
  adding, 68, 69, 212
  algebra, 77–79
  defined, 67–68
  dividing, 71, 212
  multiplying, 69–71, 212
  nut mixture problems, 137–40, 218
  subtracting, 68–69, 70

percentages
  business math applications, 183–205
  changes, 59–62, 210–12
  conversion from decimals, 56, 210–12
  conversion from fractions, 56–59, 210
  distribution, 62–65, 210–12
  tipping, 65–66
perimeters of rectangles, 100–102, 214–15
personal finances, 166–82
pi (π), 108–11, 215
powers, 83–85, 87–88, 213–14
profit
  as percentage of investment, 201–5, 225–26
  as percentage of sales, 199–201, 203–4, 225–26
proportions, 91–94, 214

quadrillions, 158–59
quantity discounts, 191–93, 225–26
quick arithmetic, 2
quintillions, 158–59

rate = distance/time, 112–20, 124–26, 216–17
ratios, 89–90, 214
real estate
  commissions, 183–86, 225–26
  tax, 176–79
reciprocals, 48
rectangles
  area of, 95–100, 214–15
  perimeter of, 100–102, 214–15
rounding, 21
rule of 70, 146–48, 218–20

sales tax problem, 170–72, 225–26
Selby, Peter, 227
speed limit problems, 126–30, 216–17
square roots, 85–88, 213–14
squares
  area of, 95–98, 214–15
  perimeter of, 101, 102, 214–15
statistics
  mean, 163–65, 222
  median, 163–65, 222
  mode, 163–65, 222
subtraction
  fractions, 44–45
  isolation of *x*, the unknown, 73–74
  negative numbers, 68–69

taxes
  federal income, 176–79, 225–26
  real estate, 176–79
  sales, 170–72, 225–26
thousands, 153–55, 221–22
time = distance/rate, 120–26, 216–17
tipping, 65–66
triangle, areas of, 102–4
trillions, 155–58, 221–22
2/10 n/30 discounts, 193–97

word problems
  advanced, 51–55
  division, 25–27
  multiplication, 17–18